Agriculture as a Producer and Consumer of Energy

AAAS Selected Symposia Series

Published by Westview Press, Inc.
5500 Central Avenue, Boulder, Colorado

for the

American Association for the Advancement of Science
1776 Massachusetts Avenue, N.W., Washington, D.C.

Agriculture as a Producer and Consumer of Energy

Edited by William Lockeretz

AAAS Selected Symposium **78**

AAAS Selected Symposia Series

This book is based on a symposium which was held at the 1981 AAAS National Annual Meeting in Toronto, Ontario, January 3-8. The symposium was sponsored by AAAS Section O (Agriculture).

Published in 1982 in the United States of America by
Westview Press, Inc.
5500 Central Avenue
Boulder, Colorado 80301
Frederick A. Praeger, President and Publisher

Library of Congress Cataloging in Publication Data
Main entry under title:
Agriculture as a producer and consumer of energy.
(AAAS selected symposium ; 78)
"Based on a symposium which was held at the 1981 AAAS National Annual Meeting in Toronto, Ontario, January 3-8"--Verso of t.p.
Includes bibliographical references and index.
1. Agriculture and energy--Congresses. I. Lockeretz, William. II. American Association for the Advancement of Science. III. Series.
S494.5.E5A39 333.79 82-2021
ISBN 0-86531-381-4 AACR2

Printed and bound in the United States of America

About the Book

The deteriorating energy situation has affected U.S. farmers in two widely divergent ways. On one hand, the sky-rocketing costs of fertilizers, equipment operation, and most other necessities have created severe problems. On the other hand, the demand for energy produced from farm products means a stronger and more diversified market for those products.

The authors of this book look at both direct and indirect effects on farmers of the changing energy situation, explore conservation methods and alternative energy sources suitable for farms, and analyze the costs and benefits of three agricultural energy sources--methane from livestock manures, alcohol from carbohydrate crops, and the energy products of various crop residues. Unlike the standard treatment of these topics, the emphasis here is primarily on how farmers--and agriculture generally--will be affected by changes in energy production, costs, and consumption patterns, rather than simply on how much surplus energy the farm sector can provide for the rest of the economy.

About the Series

The *AAAS Selected Symposia Series* was begun in 1977 to provide a means for more permanently recording and more widely disseminating some of the valuable material which is discussed at the AAAS Annual National Meetings. The volumes in this *Series* are based on symposia held at the Meetings which address topics of current and continuing significance, both within and among the sciences, and in the areas in which science and technology impact on public policy. The *Series* format is designed to provide for rapid dissemination of information, so the papers are not typeset but are reproduced directly from the camera-copy submitted by the authors. The papers are organized and edited by the symposium arrangers who then become the editors of the various volumes. Most papers published in this *Series* are original contributions which have not been previously published, although in some cases additional papers from other sources have been added by an editor to provide a more comprehensive view of a particular topic. Symposia may be reports of new research or reviews of established work, particularly work of an interdisciplinary nature, since the AAAS Annual Meetings typically embrace the full range of the sciences and their societal implications.

WILLIAM D. CAREY
Executive Officer
American Association for the Advancement of Science

Contents

About the Editor and Authors

William Lockeretz *is a research associate in the School of Nutrition at Tufts University and a specialist on energy use in agriculture and use of agricultural biomass in energy generation. He is a member of the editorial board of* Energy in Agriculture *and editor of* Agriculture and Energy *(Academic, 1977).*

Roger Blobaum, *president of Roger Blobaum Associates, Washington, D.C., served as principal investigator of the Small Farm Energy Project, a national research and demonstration project. He has served on numerous national advisory panels and committees, and he has published on farmland preservation, rural energy planning, renewable energy applications, and the use of municipal waste on agricultural lands.*

Otto C. Doering III, *a specialist in public policy and resource economics, is a professor in the Department of Agricultural Economics at Purdue University. He has written extensively on energy use in agriculture, and his work has earned him the Distinguished Policy Contribution Award (American Association of Agricultural Economics, 1978).*

Folke Dovring, *a professor of agricultural economics at the University of Illinois, is a specialist in resource and institutional economics and economic development. He has served as a consultant to national and international agencies and is the author of* Land and Labour in Europe in the 20th Century: A Comparative Survey of Recent Agrarian History, 3rd ed. *(Netherlands: M. Nijhoff, 1965).*

W.E. Larson *is a soil scientist in the Science and Education Administration, U.S. Department of Agriculture, and is primarily concerned with soil management and soil physics. A former president of the Soil Science Society of America, he has written on the use of sewage wastes on land and on the effects of tillage and crop residue removal on erosion, runoff, and plant nutrients.*

John H. Martin, Jr., *a research associate in the Department of Agricultural Engineering at Cornell University, has specialized in the utilization of agricultural wastes and in the environmental problems associated with agricultural production.*

F.J. Pierce, *an associate scientist in soil science at the University of Minnesota, is currently concerned with bioenergy, soil chemistry, and soil management.*

J.B. Swan, *a specialist in soil management and soil physics, is professor of soil science at the University of Minnesota. He has published on dryland farming, irrigation, tillage and soil management practices, and soil and water conservation.*

William Lockeretz

1. Introduction: The Agriculture-Energy Relationship

Agriculture is the pre-eminent renewable energy system. For several millenia, farmers who never heard of terms like "solar energy" have used the sun for the most fundamental of all production processes. In earlier times, energy was not just an input, but also one of the products of agriculture. Indeed, in many parts of the world livestock manure is still used as a heating fuel, and draft animals fed on farm-produced feeds are important for transportation and stationary power.

During the fossil fuel era, agriculture has served mainly to transform energy (both from the sun and from fuels) into food, fiber, and other products, rather than being an energy producer in its own right. Thus there might not appear to be any particular reason to give energy a special status when analyzing an agricultural system. On the input side, it is just one of several production resources, along with other physical ones like land, water, and labor, as well as less tangible ones like technical knowledge. Even when the input energy is purchased, rather than being available free from the sun, its cost is not necessarily significant. In fact, during most of the several decades in which purchased energy has been widely used in United States agriculture, its cost was a fairly inconsequential item in a typical farm budget.

Nor should any special significance be attached to the fact that agricultural production may be measured in energy terms, i.e., calories of food energy. To be sure, this permits the calculation of a dimensionless output/input "energy efficiency" ratio, but such a ratio at best is of rather abstract interest, given that food is not raised simply for its energy value. In fact, it is not raised at all for its energy value. It may be raised for its _nutritive_ energy value, among other things, but since the input energy--fossil fuel, say--does not have nutritive energy value, the ratio energy out/energy in is no more fundamentals nor "pure" than a ratio like protein out/energy in, or heads of cabbage out/energy in.

The recent energy situation has begun to change the status of energy, giving it a special position among both the inputs and outputs of agriculture. Sharp price increases have made energy a significant part of total production costs in some particularly high-energy production methods like irrigation from a deep well with a high-pressure distribution system. Moreover, since energy price increases have been so sudden, and since further increases are expected, more attention is being paid to energy than might seem called for purely on the basis of its numerical share of total production costs. Nor is this inappropriate. When prices of various inputs have remained fairly constant in relation to each other, then presumably farmers' practices reflect these relative prices, and the most expensive ones are already being used particularly carefully and efficiently. But since it takes time for farmers to adjust their practices, when one input's price has risen very suddenly that is the one whose use is now most out of line with its current price. Also, besides anticipated price increases, there is serious concern over supply cutoffs and unanticipated shortages. Even a brief interruption could be a serious problem for agriculture, since the timing of many energy-using agricultural operations is very critical. In several ways, therefore, the energy consumption of agriculture indeed deserves major consideration.

The second reason that energy is acquiring special importance is that agriculture once again may be called upon to be a supplier of energy, rather than only a consumer. There are several energy production systems that have long been technically feasible but that were considered economically uncompetitive. Some of these are now receiving renewed research and development attention, and one is already commercial on a significant scale. Examples include methane production from livestock manures and other wastes, gasification of crop residues and agricultural processing wastes, and alcohol from starch and sugar crops. The last is sold as a motor fuel (blended with gasoline) in several thousand service stations.

Thus whether agriculture is examined as a producer of critically important products or as a consumer of critical resources, it clearly has a fundamental relationship with the entire energy system. This book explores several parts of this very complex relationship. It does not try to cover this subject in a systematic and comprehensive way. Rather, each chapter is a self-contained presentation of one particular aspect of the agriculture-energy connection, although the authors do refer to each other whenever they can do so naturally and without stretching a point just for the sake of integrating their contributions.

The one unifying element is that the emphasis is more on what the energy situation means for agriculture, rather than what agriculture might mean for the energy situation. For example, if one's point of view is the energy sector, the relevant issues regarding agriculture's energy consumption would be how it compares to total national consumption and how much it can be cut. But from the viewpoint of agriculture, more appropriate questions would be how much energy now contributes to production costs, how much the alternatives would cost, and whether each alternative is attractive or unattractive to a farmer. The two sets of questions are related, of course, but they are quite distinct. The two chapters (2 and 3) that deal with agriculture's consumption of energy tend to take more of the latter approach.

Likewise, both viewpoints are possible when analyzing agriculture as an energy producer. One could ask how much energy agriculture could provide to the rest of the economy and how much it would cost compared to other sources. But one could also ask about its effects on the use of agricultural resources, whether it is compatible with agriculture's primary role as a provider of traditional food and fiber products, and how it will affect farm income and returns to agricultural capital and labor. Questions such as these are the underlying theme of Chapters 4 through 7, all of which deal with production of fuels from agricultural raw materials for use by the rest of the economy.

Perhaps because the energy situation has been so highly publicized and because far more people are energy consumers than food producers, most of the discussion of the agriculture-energy relationship has been from the viewpoint of the energy sector. The basic question is "How much can agriculture do for the rest of us by consuming less energy and producing more?" But a few simple numerical examples will show that some changes that would have inconsequential effects on the national energy picture could have profound effects--either beneficial or adverse--on the overall agricultural situation. Given agriculture's importance in the national economy, it follows that the agriculture-energy interaction must be explored from both directions.

For example, the energy consumed in primary agricultural production--that is, up to the point at which the output leaves the farm--has been estimated at 2 quads (2×10^{15} BTU, or 2×10^{18} joules), if we include the energy needed for fertilizer and pesticide manufacture. This is only about 3% of the national total. Thus even a very aggressive conservation program, which might, speaking optimistically, reduce this figure by half, would reduce national consumption by less than 2%. But in 1980, farmers' energy requirements cost about $14 billion. Reducing this by half (and ignoring the cost of the conservation measures, which could be considerable) results in a saving of about $7 billion, which would add about 30% to the estimated $24 billion net income from agriculture in 1980.

Likewise, a fairly ambitious alcohol fuel program might yield some 5 billion gallons (19 billion liters) annually, which would replace only about 5% of current gasoline use. But it would require about 2 billion bushels (50 million metric tons) of corn, assuming it all was produced from corn. (This is not a very oversimplified assumption for the near term, as will be discussed in Chapter 5.) This is more than a quarter of the entire crop. Moreover, the by-product livestock feeds produced along with the alcohol (discussed further in Chapter 4) would have the protein equivalent of two-thirds of the current consumption of soybean meal, the main high-protein livestock feed. Since corn and soybeans are the nation's first and second biggest crops respectively, such an alcohol program clearly would have major implications for the entire demand and supply structure of American agricultural markets, even while having only a rather minor effect on the overall liquid fuel situation.

These examples of direct and obvious effects only begin to reveal the enormous potential significance that the changing energy situation has for the nation's agriculture. Other effects are more subtle and indirect, but no less significant. Agricultural exports have been very much in the news in recent years. The most dramatic example was the 1980 embargo on grain exports to the Soviet Union because of its actions in Afghanistan. There also have been real or threatened embargoes on exports to our allies for economic reasons, such as an earlier short-lived embargo on soybean sales to Japan that was strongly resented by that country. The sudden surge in grain exports to the Soviet Union in 1972, an important factor in the sharp crop price increases of the early 1970s, became a major domestic political issue. Moreover, the record high value of agricultural exports, currently around $40 billion annually, is obviously an important factor when oil imports alone are near $100 billion. Since soybeans and corn are respectively our two leading export crops, major changes in their supply and demand because of an alcohol fuel program would be important not only domestically but also with regard to our agricultural trade relations, an issue explored in Chapter 4.

Agriculture's production or consumption of energy can also have important effects on other finite resources. For example, crop residues are regarded as an attractive energy raw material because they are not also usable as food. But they are very important in preventing soil erosion and in preserving the soil's fertility. As shown in Chapter 6, a significant fraction, but only a fraction, of the total residue production can be removed from the soil without serious damage. If economic forces cause farmers to sell additional residues beyond what is compatible with soil constraints, the long-range effect on agricultural productivity could be far more important than the relatively minor contribution that the additional residues could make to the national energy supply. Moreover, soil erosion causes an environmental problem besides the loss of an important resource. Soil eroded by water contributes to pollution of surface waters by sediments, pesticides, and nutrients, while soil eroded by wind in one area becomes an air pollution problem somewhere else. While a renewable alternative to fossil fuels would help reduce some serious problems, in balance the environmental effect could still be adverse if agriculture is allowed, or compelled by economic factors, to disregard its own resource limitations and its potential for environmental damage. On the other hand, some energy production from agriculture--most notably methane from livestock manures, discussed in Chapter 7--can offer some help in reducing current agricultural environmental problems.

Finally, agriculture's response to the energy situation could affect farmers' economic well-being in ways that are not reflected in simple cost and income accounts, as is discussed in Chapters 2, 3, and 4. Here too, the effects can be either beneficial or harmful, depending on the particular approaches taken and on one's goals and values. Thus a renewable energy system might be a good use for a farmer's surplus labor and management ability during non-peak seasons, but could also be a serious burden when the most pressing need is to plant or harvest the crop on time. It is notoriously difficult to attach a specific dollar value to a farmer's own labor, which greatly complicates a conventional economic analysis of an on-farm energy system.

Besides the direct and indirect economic effects that occur at the level of the individual farm, consumption and production of energy have major implications for a set of larger-scale issues of agricultural structure. Changing energy supplies and rising prices will have different effects on various farms according to their size, location, specialization, and current use of inputs, as discussed in Chapter 2. Renewable energy systems, such as the solar collectors described in Chapter 3, will give farmers a greater degree of self-sufficiency with respect to input suppliers, and therefore will have structural consequences, even if their costs just offset the savings in fossil fuel. On-farm renewable energy systems that use agricultural products as the basic raw material, most notably alcohol production from grain, will have particularly important structural implications in that they provide farmers with a new outlet for their production, and therefore offer some marketing flexibility and leverage against grain buyers, as is pointed out in Chapter 4. It may turn out that the most attractive feature of on-farm energy production is not that it actually saves money directly--which indeed it may not, given its costs--but rather that a larger fraction of the total value of agricultural production is retained on the farm, or at least in the local economy. This could occur either because resources already available on the farm replace energy or energy-intensive inputs that currently are purchased, or because additional value beyond that of primary production is added by converting crops or residues into saleable fuels. A farm that achieves a given net income through a more diversified enterprise mix, or with a lower cost/revenue ratio, is less vulnerable to decreases in the prices of specific agricultural products or to increases in the prices of inputs such as fuel.

In summary, energy is important for agriculture in ways that are far-reaching, complex, and sometimes quite indirect. Proposals concerning how agriculture should respond to the changing energy situation do not have much hope for offering an acceptable balance between the interests of the producers and consumers of either energy or food unless they reflect the pervasive role of the agriculture-energy interaction.

Fortunately, the scope of public concerns regarding agriculture has broadened significantly in recent years, and the scope of agricultural research has broadened accordingly. Some deeply held but simplistic notions regarding what constitutes efficiency, equity, and progress in agriculture have at least been undergoing critical reexamination, even if they haven't been rejected outright. The catch phrase "get big or get out" has given way to serious public discussion of what agricultural structure we want for the future, such as whether we should expressly intervene to preserve smaller farms. Instead of a virtually unquestioned acceptance of the need to reduce agricultural labor--unquestioned, that is, by people other than those whose labor was no longer needed--we have seen prominent agricultural officials question whether public funds should be spent for research on labor-displacing mechanization. We also have come to acknowledge the fundamental difference between renewable and depletable production inputs--not just energy, but also irrigation water, for example--a difference that has implications going far beyond what the farmer pays for them. And whereas the approach during most of the post-World War II period generally was to try to deal with the "farm problem" through rather narrow price support and production control programs concerned mainly with particular commodities, more recent farm legislation has explicitly dealt with consumer interests, resource conservation, structural factors, and research priorities.

This admirable broadening of the range of publicly discussed agricultural issues in the 1970s may not yet have resulted in many tangible changes, and indeed never will if it turns out to have been a temporary aberration and is not followed by funds, programs, and commitments to turn it into something more than words. But at least it means that new problems, such as the changing energy situation, will be analyzed in a suitably broad context, and that discussions of appropriate solutions will give adequate attention not only to purely technical and micro-economic factors, but also to related resource, environmental, and structural issues. This book is offered as a contribution to such a discussion.

Otto C. Doering III

2. The Effects of Rising Energy Prices on Agricultural Production Systems

In assessing the potential effects of rising energy prices on agricultural production systems it is first necessary to define more clearly what is meant by rising energy prices. Rising prices are only meaningful as a relative concept. If all prices rise at the same rate, consumers are not given a reason to change their preferences for comparable goods. In agriculture, as in other production processes, changes in the relative use of different resources will often occur when there is a shift in the price of one relative to another, given that one can often serve as a substitute for the other.

Inputs such as land, labor, and fertilizer are the ones most commonly thought of in connection with the agricultural production process. A different kind of input that is especially important to this discussion is technology. Technology can be thought of as an input in its own right that substitutes for some other input, or it can be thought of as something which allows or facilitates the substitution of less costly for more costly resources. An example of the former is a new machine that substitutes for labor. An example of the latter is a new practice or technique which may have no explicit cost but which brings about increased production from existing inputs or improves their effectiveness.

Over the past several decades, virtually all farm input prices have increased in absolute terms. But some of these inputs have had rather rapid price increases, while increases in others have been moderate. The pattern of price increases for several critical inputs is shown in Table 1. Table 1 shows a trend of declining relative prices for gasoline and fertilizer as compared with land and labor from 1950 to 1970. Since 1970 there have been large price increases in petroleum products, so that gasoline and fertilizer have increased more in relative price than machinery or labor. The trend of

Table 1. Prices Paid for Selected Farm Inputs in the United States

Year	Index: 1950 = 100				
	Farm Wages	Machinery	Farmland	Fertilizer	Gasoline
1950	100	100	100	100	100
1955	121	113	131	108	112
1960	148	138	171	106	119
1965	171	154	214	106	123
1970	255	191	294	90	135
1975	382	323	539	224	204
1980	565	525	1,004	255	450

Source: USDA (1951, 1956, 1961, 1967, 1969 and 1980); ESS (1981).

constantly declining relative prices for petroleum products and fertilizer has been broken. However, it will take further large market price increases in petroleum and fertilizer coupled with stable or declining land prices, wage rates and machinery costs to effectively alter the relative input price relationship of the 1950s and 1960s. This price relationship encouraged the development of the high energy -- high fertilizer system of agricultural production that we have today. Future changes in agriculture will depend on the relationship between energy and other input prices, and in particular on whether there is a sustained shift in this relationship. Such a shift might take a decade to occur, after which there would be a comparable or greater lag while the agricultural production system as a whole adjusted to it.

Farm-Level and Structural Effects of Rising Energy Prices

The effects of energy price increases will be felt on several levels. Besides causing adjustments on individual farms, price increases will also affect the larger structure of the agricultural production sector, which is made up of many farm firms. "Structure", a broad term referring to ownership, control, size and concentration of farms, has been the subject of considerable public debate recently. Former Secretary of Agriculture Bergland was especially concerned with the increasing concentration of ownership and control and the increased size of those farms that produce the bulk of the food and fiber in the United States. A recent USDA study noted that a decreasing proportion of the nation's farms produces an increasing proportion of the food and fiber, and that only five percent of the nation's farm owners control almost half of the farmland (USDA, 1981). The concern here is not whether this past trend is good or bad, since this will depend upon an individual's values. We want to know what impact rising energy prices might have on current structure or on the trend that has developed over the past several decades.

Many factors affect the structure of agricultural production. Some of these are: demography, climate, the resource base, input costs, control of inputs, technology, the structure and control of processing and markets, control and availability of transportation, transportation costs, consumer demand, and finally, government policy. Energy prices are a factor in only some of these, and in many cases play only a minor role. In the overall food and fiber sector the value of direct energy inputs has been only a small fraction of total product value as indicated in Table 2. These figures indicate that even substantial energy price increases would provide only relatively moderate inducements to change the production process, let alone the structure of the production sector.

Table 2. Value of Direct Energy Input as a Percentage of Total Product Value at Each Stage of Production

Sector	Percentage of Total Product Value
Food Production	5%
Food Processing	1%
Food Marketing	3%
Food Preparation at home	7%
Food Preparation away from home	12%

Source: Van Arsdall and Devlin (1978); Doering et al. (1977).

A similar relation exists between direct energy costs and other production costs. Gasoline and diesel fuel are what generally come to mind in connection with energy use in agriculture. The cost of all fuels and lubricating oils used to farm an acre in the Corn Belt over the entire growing season, including harvest, comes to about $15 to $20. By comparison, cash rents up to $150 per acre are not uncommon in the region. Moreover, other less obvious energy costs are more significant in dollar and energy terms. Compared to the roughly 10 gallons of gasoline energy (or its diesel equivalent) used per acre for field operations in corn production, the drying of grain from that acre requires 20 gallons of gasoline energy equivalent, while manufacturing the fertilizer requires 40 gallons equivalent. In the case of irrigated corn, if deep wells are required and the climate is dry, the energy for irrigation can be two to three times as large as the energy for fertilizers, or an order of magnitude greater than for field operations.

The obvious, direct effects of rising energy prices are only part of the story. There will also be indirect effects. The remainder of this paper examines both kinds of effects from some major farm energy inputs and considers how they might affect production systems and ultimately farm structure.

Direct Effects

Fertilizer Costs

Rising energy prices can certainly increase the cost of some important farm inputs, especially fertilizer. The abundant supply and relatively low price of chemical fertilizer, especially nitrogen, has been a major factor in the productivity and structure of our current agricultural system.

Anhydrous ammonia, one of the most common forms of manufactured nitrogen fertilizer, currently costs about $200/ton. The main energy component in anhydrous ammonia is natural gas, which is used both as a feedstock and for some process heat. The industry uses about 38 thousand cubic feet (MCF) of natural gas per ton of anhydrous ammonia, and in the past the cost of this natural gas has been very low.

The current cost of natural gas to the fertilizer industry is about $2.00 to $2.50/MCF as a result of recent federal actions to decontrol natural gas prices eventually. At a price of $2.30/MCF, the cost of the natural gas component in a ton of anhydrous ammonia is approximately $87.

Under full price decontrol the price of natural gas to the petrochemical industry might be around $5/MCF. One would not expect a greater price increase in the intermediate term because there would be some increase in supplies, based on our experience over the past year with partial decontrol. In addition, some large classes of natural gas users, like households, will have strong economic incentives to invest in conservation measures like extensive home insulation, which would slow demand. However, let us assume a more extreme case, namely that government policy is to set the minimum natural gas price at $7/MCF, which is equivalent (on an energy basis) to diesel fuel at about $1/gallon. This new natural gas price would result in a price of $380/ton for anhydrous ammonia. A tripling of the natural gas price results in less than a doubling of the fertilizer price because the other costs of fertilizer production, like capital investment, are held constant for this example. Two relevant questions are what this price increase means in terms of a farm firm adjustment, and whether this results in structural change. In looking at the first question the critical factor is what price level is assumed for the output, i.e., the corn that is produced. We have recently been through a period when corn sold for around $2.50/bushel, while the price of anhydrous ammonia was about $200/ton or 12¢ per pound of N. (Ammonia is 82% N.) We are entering a period when it appears as though corn might sell for $3.50/bushel, which we can compare to a price of 23¢/pound for N (equivalent to $380/ton of anhydrous ammonia). These price relationships then need to be analyzed on the basis of a production function for corn that relates the additional amounts of nitrogen applied to the resulting additional yield.

Figure 1 illustrates such a relationship. The data presented here are somewhat dated and thus very conservative by today's experimental results. In fact, Figure 1 shows less response to nitrogen at higher fertilization levels than is obtained by many Corn Belt farmers today.

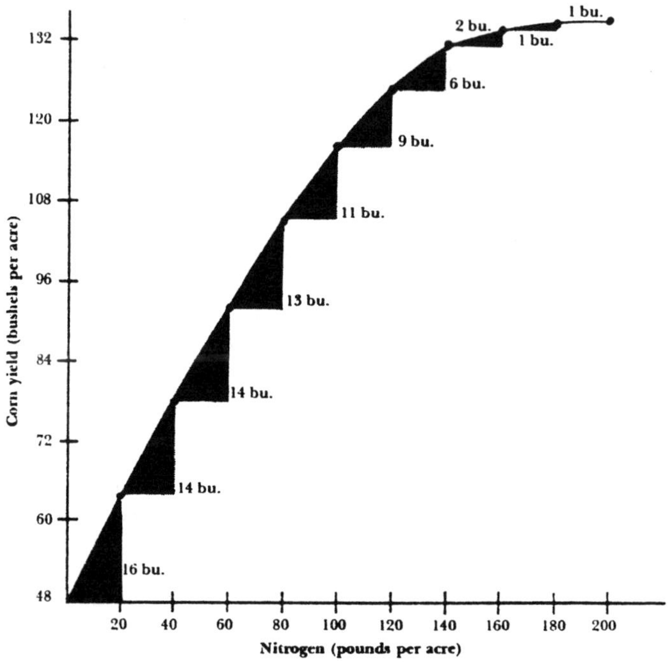

Figure 1. Corn response to nitrogen on Odell silt loam in central Indiana, 1967-69 (from data reported by R.K. Stivers, Agronomy Department, Purdue University).

Table 3 uses the data from Figure 1 to compare the cost of additional units of fertilizer to the value of the additional grain produced with that fertilizer, for both of the fertilizer/corn price ratios mentioned earlier. The table shows that the breakeven point between the cost of an additional unit of fertilizer and the corn produced from that unit is around 180 pounds of N for both fertilizer corn price ratios. Our actual experience has been that over the last three years corn farmers in Indiana applied N at an average of 140 pounds/acre. Even with corn remaining at $2.50 while the cost of N increases to 23¢, the breakeven point is just above 160 pounds of N.

Therefore, even a very large price increase in natural gas would probably not result by itself in a decrease in nitrogen fertilizer use if commodity prices follow current trends. Even with low commodity prices, nitrogen use would be reduced only slightly, if at all.

If there were a severe reduction in the use of manufactured nitrogen, this would have a definite impact on production systems and the structure of agricultural production, if farmers responded by using alternative sources rather than just using less. The replacement of chemical nitrogen with legumes would encourage mixed farming systems and require more land for a given amount of cropland under cultivation. However, over the next decade energy prices are not likely to go high enough to raise the price of nitrogen fertilizer and therefore the fertilizer/corn price ratio by enough to reduce substantially the use of manufactured nitrogen.

Irrigation Costs

Low cost energy has been an important factor in the development and extension of irrigation for food and fiber crops. In areas like the Western Corn Belt and the Great Plains, irrigation has been an essential component of agricultural development and a major determinant of the current structure of agricultural production. Increased energy costs are already causing shifts in cropping patterns that should result in structural change.

It is necessary to make a distinction between different types of irrigation. Surface irrigation based on gravity feed will hardly be affected by higher energy prices, except as water may be withdrawn from agriculture to generate power. Low pressure pump systems with gravity distribution will also suffer little impact. High energy systems such as overhead spray irrigation from deep wells will be severely hurt by higher energy prices, and are beginning to suffer already.

Table 3. Comparative Returns with Different Fertilizer/Corn Price Ratios

N Application (lb/A)	Yield (Bu/A)	N @ 12¢/lb; Corn @ \$2.50/bu		N @ 23¢/lb; Corn @ \$3.50/bu	
		Added N Cost	Added Corn Returns	Added N Cost	Added Corn Returns
60	92	—	—	—	—
80	105	\$2.40	\$32.50	\$4.60	\$45.50
100	116	\$2.40	\$27.50	\$4.60	\$38.50
120	125	\$2.40	\$22.50	\$4.60	\$31.50
140	131	\$2.40	\$15.00	\$4.60	\$21.00
160	133	\$2.40	\$ 5.00	\$4.60	\$ 7.00
180	134	\$2.40	\$ 2.50	\$4.60	\$ 3.50
200	135	\$2.40	\$ 2.50	\$4.60	\$ 3.50

Table 4. Energy Costs for 1 Acre of Corn Production, Oklahoma High Plains[a]

Price of Natural Gas ($/MCF)	Cost of 200 lb of N[b] ($/A)	Cost of Energy for Water Pumping ($/A)
$2.30	$24	$ 32
$7.14	$47	$100

[a] Source: Eidman, Dobbins and Schwartz (1975). Based on 1½ - 2 acre feet of water, pumped from 300-400 feet.

[b] Based on 38 MCF of natural gas per ton of anhydrous ammonia with non-energy costs of $113/ton.

One area where overhead spray irrigation is particularly important is the Oklahoma High Plains. It is interesting to compare the effect of natural gas price increases on the cost of irrigation water from deep wells with their effect on the cost of fertilizer. As in the previous fertilizer example, assume an initial natural gas price of $2.30/MCF, and compare this with a price of $7.14/MCF (which is equivalent to diesel fuel at $1/gallon). Assume that nitrogen is applied at 200 pounds/acre and that natural gas supplies the energy needed to meet the water requirements of the irrigated corn crop.

Cost increases have already been staggering. Today, farmers are paying ten times as much for natural gas as they paid a few years ago. Over this longer time period, the cost of energy for pumping goes from being a small fraction of the fertilizer bill to twice the fertilizer bill under an extreme pricing policy in the future. As a result of increased energy costs, it is expected that much of the irrigation in areas like the High Plains will be discontinued by the year 2000 (Young and Coomer, 1980).

In examining the potential structural effect of higher irrigation energy costs, it is worthwhile to recollect the historical development of the Western Corn Belt and Great Plains states. These areas started with small holdings operated under unusually favorable (wet) weather conditions over much of the period 1890 to 1920. Upon a return to normal dry weather the Dust Bowl occurred and there was a trend, encouraged by government policy, towards farm consolidation and moisture conserving technology such as summer fallow. The onset of modern irrigation technology led to further farm consolidation, since those having the capital to invest in such equipment could buy out their neighbors' farms. If irrigation must now stop in some of these areas, it will mean a return to more extensive dryland farming with moisture conserving techniques. More extensive holdings will be required to produce the same income, and land values will have to decline to reflect the lower productivity of this land. This presents a potential equity problem for those who recently purchased this land at prices reflecting its irrigated potential at low energy costs. It also presents us with a substantial drop in the productive capacity of what is today some of the nation's most productive farmland.

Indirect Effects: Transportation Costs

Cheap transportation has been critical to the development of the current structure of U.S. agriculture, especially to its high degree of regional specialization. Because the cost of transport to market has been low, the location of production has been largely determined by the most favorable climate and the best situation for other production resources. Thus, production of processing tomatoes has been concentrated in California -- the New Jersey Tomato has died and the Indiana Tomato has just survived.

The energy used in transportation in the whole food and fiber system has been relatively modest in absolute terms (Barton, 1980). Public investments like the interstate highway system have also kept transportation costs down. Today, higher energy costs are pushing freight rates up and the public is increasingly concerned that roads, waterways, and other public transportation networks should pay their own way. The increase in truck freight rates after 1978 was sufficient to change the economically preferred location in several instances. As an example, it became more profitable to fatten Kentucky calves in Kentucky, rather than ship them to Texas feedlots to be fed on irrigated corn or corn shipped in from the Corn Belt.

Higher energy costs, and the resulting higher transportation costs (especially for truck transport) should lead to a relocation of food and fiber production and to a rationalization of the mode of transportation. The first of these would have a major impact on the structure of production -- tomatoes for processing would be grown in several areas of the country, not just in California. The second would result in little locational or structural change in the agricultural production system. As an example, the recent truck strike encouraged a major railroad to attempt to regain some of the shipping business for fresh California produce that the railroads held before the interstate road system and the post World War II Interstate Commerce Commission regulations. On the basis of this modest and successful experiment, fresh produce may continue to be grown in California because of the extremely favorable climate and because this produce will be shipped on energy efficient trains meeting the rigid time schedules necessary to avoid spoilage. The railroads were previously able to meet such schedules regularly in the early 1900s, and there is no reason why they shouldn't be able to do so today.

The location of production is closely linked to that of processing and marketing. Thus, the energy economics of production may not be the primary factor in location. Existing processing and marketing arrangements may be sufficiently favorable that higher transportation costs will be absorbed. Conversely, shifts in energy costs or in other economic factors in processing and marketing may even move production to less energy efficient areas. The production system or its location is often determined by factors outside the production process itself. Total system efficiencies usually prevail and are thus an essential factor in any analysis.

Joint Impacts Affecting Structural Change: The Case of Inflation

Changes in structure often result from several long-term influences that combine to overcome some opposing influences. It is usually impossible to pinpoint one factor as the cause for a given structural change. Because of this it is important to see how factors other than energy that are critical by themselves to structural changes in agricultural production may also interact with rising energy prices.

Farming has traditionally been an industry with very low returns to total invested capital, primarily because of the very high long-term investment in land. Investment per worker is much larger in farming than in industry or manufacturing. Since World War II returns to land have been in the range of 2 to 5%. Because of the large number of individual farm firms, farmers are price takers and have no market power that would allow them to increase commodity prices as costs increased, and thereby maintain or improve their returns to land.

Consequently, inflation has had an especially severe impact upon individual farms and has also become a major factor in structural change in the agricultural production sector (Guither and Doering, 1980). In order to illustrate the impact of inflation we can look at the purchase and operation of a farm when inflation is at 9% compared to when it is zero. In both cases we will assume that the farm is purchased entirely on borrowed money at the interest rates that could be expected with 9% or no inflation (12% and 3% respectively).

Table 5. Representative Farm Returns and Cash Flow (% of Farm Value) with Full Debt Financing of Land Cost

	Inflation Rate	
	0%	9%
Returns		
Current Earnings	4%	4%
Deferred Earnings	0%	9%
Total Earnings	4%	13%
Current Cash Flow		
Current Earnings	4%	4%
Mortgage Rate	3%	12%
Current Surplus (Deficit)	1%	(8%)

Current earnings remain the same in both cases, and are expressed as a return on the value of the farm. With 9% inflation, the value of the farmland increases over time at a rate equal to the rate of inflation. However, this is not a current return and is only realized later, when the farmer sells his farm.

Note the current cash flow deficit in times of inflation as compared with the period of no inflation. Inflation forces the farm operator to defer earnings in the form of long term capital gains on the sale of the farmland, whether or not that is what he desires. Inflation in this case imposes a cash flow deficit for those who must finance the purchase of basic farm assets through debt during the inflationary periods.

The impact on a farm business of a factor such as inflation is far greater than that of energy costs. If these inflation-related cash flow squeezes are prolonged, it will not be possible for some people to own farmland. To be able to purchase farmland during such periods, a farmer would have to have enough of a farm business already to be able to stand the current cash flow deficit. Alternatively, the farmer might hold an off-farm job and so could "carry" the farming operation with outside income. We might see large farms getting larger through the purchase of additional land, while part-time farmers become more numerous by supporting farmland purchases with outside income. The operator of the moderate sized family farm would be squeezed. This would be a farm just large enough to support a family with income from the farm, but with little cash flow left over to support the purchase of additional land. In addition to the part-time farmer and the large farmer, outside investors willing to cover cash flow, defer income and wanting a hedge against inflation would be likely purchasers of farmland during inflationary times. Deflation would have similar perverse effects which would have great impact on the farm operation.

In addition to the direct cash flow consequences of inflation there are likely to be some more indirect or joint effects. As an example of this, rising energy prices together with an inflation-induced cash flow squeeze might result in a decrease in farm productivity. Returning to the example of the comparative fertilizer returns on Table 3, it is conceivable that a farmer might be in a situation where increased fertilizer costs coupled with a severe current cash flow problem would lead to levels of fertilizer application well below the breakeven point.

Capitalizing the Energy Efficiency of Existing Factors of Production

Another question, as energy prices rise, is the extent to which inherent energy efficiency will be reflected in the costs of other farm inputs. As an example, will there be a much greater price differential between fertile land with adequate rainfall in the Corn Belt as compared with land that may need irrigation or is less responsive to chemical fertilizer? As this occurs it will give an advantage to those already holding the more favorable resources, while those holding less favorable resources will suffer a comparative loss. On the other hand, factors other than income generating potential, such as an inflationary psychology, also help determine the price of land and other resources.

The Role of Technology

So far we have concentrated on factors whose use in the production process is determined by their relative price, or upon factors that change the economics of the process and thus affect resource use. There also are important factors -- management and technology -- that change the overall environment in which inputs are used. These factors may change the production process itself and cancel the effects of changing relative prices. Predicting the impact of technology or management is much more difficult and seems less exact than speculating on relative price scenarios. It is essential, but has not received as much attention as it deserves.

If rising energy prices induce changes in technology, what will be the effect of that technology on structure? New technology might give us synthetic fuels that could replace fossil fuels. It might enable us to substitute other resources for fossil fuels. The new technology might be high or low in its initial capital cost as compared with its operating cost and payback. All of these are going to affect the ability of different production systems and different types of farm firms to adopt such technology beneficially, which in turn will affect trends in ownership, control, size and concentration of farm firms.

Management and Information

Closely related to the role of technology is the management of this new technology and of the farm firm as a total unit. Some new techniques save energy by substituting improved information and a higher level of management skills. New pest management techniques are a striking example. It is clear that the costs of chemical pest management are high enough that now we can economically substitute improved information (down to detailed monitoring of micro-climates) and advanced management skills for these chemicals (Miranowski, 1980). Systems will change and these management and information requirements will affect structure insofar as certain types or sizes of firms can pay for and provide them while others may not be able to. There will be a minimum farm size with sufficient income to reimburse such a level of management. In fact, this requirement of earning enough to be able to afford a competitive level of management already has been an important factor in the adoption of new production systems and the consolidation and increased concentration of farms over the last several decades.

Summary

The nature of farm production systems and structure of the agricultural production sector are the product of many forces. These forces often operate together, or more likely in partial opposition, over long periods of time, before specific trends and causal relationships can be detected and confirmed. Energy prices are only one of these factors, and in many cases a minor one. To have a major impact, the price of energy must change relative to those of other resources, and its overall impact must be large enough to be important in the firm and/or system context. In addition, the price relationship of inputs to outputs is also critical. If energy prices do change sufficiently to be important in relative terms, then the nature of new technology developed in response to these new prices can be a critical factor in the direction and magnitude of structural shifts. Finally, factors such as managerial capacity, the cost of information, and the indirect influence of rising energy prices on the costs of other factors of production will also enter into the determination of production systems and structure.

It is commonly believed that severe energy price increases in absolute terms must result in dramatic short-run shifts in agriculture's energy use, and will thereby cause immediately visible changes in the structure of a sector that has been so affected. The conclusion of this paper is that such shifts are likely to take a decade or more before they become apparent, and that their direction is by no means certain. Some things that can be expected are: the development of new technology that has the ultimate effect of reducing petroleum and natural gas requirements; increased management and information requirements; and a very large number of adjustments in food production because of the effects of rising energy prices on the input, processing, marketing and food preparation sectors. The latter may be unexpected because we often do not link changes in production to changes in other sectors of the food and fiber system.

References

Barton, John A. 1980. Transportation Fuel Requirements in the Food and Fiber System. Agricultural Economic Report No. 444. U.S. Department of Agriculture, Economics, Statistics, and Cooperatives Service. Washington, D.C.

Doering, Otto C., Earle Gavett, Thomas Van Arsdall, Patricia Devlin, and Edward Rall. 1977. Current Energy Use in the Food and Fiber System. U.S. Department of Agriculture Briefing Paper. Economic Research Service, Washington, D.C.

Economics and Statistics Service, U.S. Department of Agriculture. 1981. Prices (for June 3, 1981). Indiana Crop and Livestock Reporting Service. Lafayette, IN.

Eidman, Vernon, Craig Dobbins, and Harold Schwartz. 1975. The Impact of Changing Energy Prices on Net Returns, Production Methods and Kilocalories of Output for Representative Irrigated Farms. Agricultural Economics Department, Oklahoma State University. Stillwater, OK.

Guither, Harold, and Otto C. Doering. 1980. The Family Farm and Other Choices: Issues Concerning the Structure of Agriculture. North Central Region Extension Publication, NCR No. 143. University of Illinois. Champaign, IL.

Miranowski, John A. 1980. Estimating the Relationship Between Pest Management and Energy Prices, and the Implications for Environmental Damage. Paper presented at the annual meetings of the American Agricultural Economics Association, University of Illinois. Champaign, IL.

United States Department of Agriculture. 1951, 1956, and 1961. Agricultural Statistics (for 1951, 1956, and 1961). Government Printing Office. Washington, D.C.

United States Department of Agriculture. 1967, 1969, 1976, and 1980. Handbook of Agricultural Charts (for 1967, 1969, 1976, and 1980). Government Printing Office. Washington, D.C.

United States Department of Agriculture. 1981. A Time to Choose: Summary Report on the Structure of Agriculture. U.S. Department of Agriculture. Washington, D.C.

Van Arsdall, Thomas, and Patricia Devlin. 1978. Energy Policies: Price Impacts on the U.S. Food System. Agricultural Economic Report No. 407. U.S. Department of Agriculture, Economics, Statistics, and Cooperatives Service. Washington, D.C.

Young, Kenneth, B. and Jerry M. Coomer. 1980. Effects of Natural Gas Price Increases on Texas High Plains Irrigation, 1976-2025. Agricultural Economic Report No. 448. U.S. Department of Agriculture, Economics, Statistics, and Cooperatives Service. Washington, D.C.

Roger Blobaum

3. Toward Energy Self-Sufficiency: The Small Farm Energy Project Experience

The concept of energy self-sufficiency for American agriculture is receiving increasing attention. Steadily rising farm fuel and electricity costs, which totaled $7.9 billion in 1979, are encouraging farmers to begin moving in that direction (USDA,1980).[1] The possibility of supply shortages or interruptions during planting and harvest times provides an added incentive.

Energy self-sufficiency is one way to reduce agriculture's vulnerability to energy shortages and price increases. Although energy price increases in the 1970s had little adverse impact on agricultural output, there is considerable concern that shortages or interruptions could result in dramatic increases in production costs and substantial production cutbacks (Buttel et al., 1980).

These possibilities influenced the Department of Agriculture's decision to support the self-sufficiency concept. Its announced goal of energy self-sufficiency for production agriculture by 1990 would be reached by conserving energy and by developing and applying alternative sources (Williams, 1979). Preliminary estimates suggest that solar, wind, and biomass energy could account for 25% of the net energy required, with energy consumption contributing the balance.

A long-range USDA research plan has identified 27 use categories where energy reduction is possible (USDA, 1979). The categories with greatest potential are irrigation,

[1]This figure does not include the energy components of the $6.7 billion spent for fertilizers and lime, the $3.1 billion for pesticides, or the $4.1 billion for machine hire, custom work, and contract labor.

tillage, crop drying, greenhouse heating, space heating of livestock and poultry buildings, and water heating for dairies. A related area with potential for savings is farm residences.

Most research dealing with on-farm sources of energy involves capital-intensive methods, such as generating methane from manure produced in confinement buildings and large feedlots. The resulting systems would be purchased from commercial suppliers, and would usually cost more than most farmers are willing to pay. There has been little publicly supported research on low-cost, owner-built systems that utilize the labor and know-how of individual farmers (Blobaum, 1980).

Most of the commercial systems being developed have not been demonstrated on working farms. The main government-initiated demonstration program began in 1978 and now involves model projects on about 90 farms (DOE, 1978). This on-going USDA effort is testing systems designed to reduce fossil fuel consumed in drying grain and other crops, heating and cooling greenhouses, and heating livestock and poultry buildings. Nearly all are commercial applications designed for larger farms where the operators are more likely to purchase energy systems than to design or build their own.

The Small Farm Energy Project

The first systematic effort to demonstrate the full range of proven energy applications on working farms was undertaken by the Small Farm Energy Project, a national research and demonstration project funded in 1976 by the Community Services Administration. This 39-month effort was designed to show how far a group of 25 farm families could progress toward energy self-sufficiency with an energy education program, individual technical assistance, and cost-sharing. This approach emphasized energy conservation, including weatherization, and site-built solar and other energy systems. Participating farmers were encouraged to reduce the quantity of energy they purchased by making their operations more energy efficient and by producing some of their own energy.

Although the technical feasibility of all the innovations adopted by the farmers had been established, their economic feasibility and their appropriateness for family-operated farms with limited capital and cash flow had not been demonstrated. The innovations included production of flow energy (solar, wind and biomass), resource recycling (efficient manure handling and production of compost from manure), and conservation of fossil fuels (weatherization and minimum tillage).

The project had six main objectives: 1) measuring the types and amounts of energy used by farm families operating diversified crop-livestock units; 2) calculating the energy and monetary effects of various innovations; 3) developing an income and energy recordkeeping system; 4) testing various approaches to providing technical assistance on energy innovations to farmers; 5) developing an energy education program suitable for farmers; and 6) testing cost-sharing as an incentive for adoption of energy innovations.

Participant Selection

Fifty fulltime operators of crop-livestock farms in Cedar County in northeast Nebraska were selected late in 1976 as project participants. All were limited-resource farmers with net incomes below 125% of the poverty level established by the federal government. None had ever been involved in an organized energy conservation program or had any experience with solar or other alternative energy systems.

Twenty-five were selected as cooperators and received technical, cost share, and other assistance over a three-year period. The other twenty-five were selected as a control group that only had to maintain detailed energy and income records. This included filling out record books provided quarterly. No energy-related assistance was provided to the control group.

Table 1. Project Participant Characteristics[a]

Average	Cooperators	Control Group
Family Members	4.96	5.00
Farm Size (Acres)	357.00	357.00
Cropland (Acres)	239.00	241.10
Row Crops (Acres)	121.00	130.80
Dairy Cows	39.00	34.30
Sows	10.00	14.80
Tractor Size (HP)	54.00	53.30
Cars	1.40	1.40

[a] These mean values apply to the 50 original participants. One was lost from each group after the project began, leaving 48 farm families that participated throughout the 39-month period.

The cooperators were selected from among 70 farmers who submitted applications after reading about the project, attending local information meetings, or receiving a mailing sent to farmers on lists obtained from two local banks and the county office of the Farmers Home Administration (FmHA). All of the applicants expressed interest in participating in an innovative energy project. Control group members were recruited from among limited-resource farmers who did not respond to project-related publicity or mailings. Neither group was a representative sample of farmers in Cedar County or any other geographical area.

An attempt was made to match the two groups on an aggregate basis. Table 1 shows, for example, that average farm size for the two groups was identical (357 acres) and that the average number of family members was about the same (4.96 for the cooperator group and 5.00 for the control group). This was important because the project covered energy used for both family living and the operation of the farm.

Energy use data for the two groups for 1976 provided further evidence that they were well matched. Farm records showed that the average consumption of purchased energy was 976 million BTUs for the cooperator farms and 978 million BTUs for the control group farms. Another indicator of a good match was the 20% increase in use of motor fuels, including gasoline and diesel fuel for tractors, for both groups of farms in 1977 as field operations were stepped-up after a three-year drought.

Constraints Affecting Participants

An important constraint for the cooperating farmers was that they did not have farm income that could be diverted from living expenses and invested in alternative energy systems. Their average gross income was $33,000 in 1976, the year they were selected, and average net income per farm was $2,600. The cooperator families spent an average of $3,986 in 1976 for purchased energy.

Consequently, the project decided to use cost sharing, a method of providing financial assistance that has been used for many years in federal farm programs. This provided some initial capital to cooperating farmers trying new energy innovations and reduced the risk involved. Federal energy tax credits did not become available until after the project was underway.

Although all of the cooperators except one owned some of the land they farmed, including the site of their homes and other buildings, they were unable to obtain financing for their energy systems from the Farmers Home Administration. This federal agency, which makes low-interest loans to limited-resource farmers, was unwilling to make loans on solar systems. Local bankers, who supported the project in the community, were willing to make loans. But they were available only at a higher interest rate than loans for livestock, machinery, and equipment.

Most of the cooperators, as a result, covered their share of the cost of solar and other systems from savings or cash flow. They invested a total of $29,700 in cash (an average of about $1,200 per farm), a considerable amount of used lumber and other recycled materials, and hundreds of hours of time in the projects. The balance of the financing came from cost-sharing payments, which ranged from 10% to 75% of the cost of purchased materials. These payments covered about 40% of the total cost of purchased project materials.

These financial constraints led the cooperators to apply strict tests of practicality and economic feasibility to each project and made them reluctant to build anything expensive. With the exception of an attached solar greenhouse, none of the site-built energy projects cost more than $2,000, and nearly all had estimated payback periods of seven years or less.

The shortage of capital limited the adoption of wind energy systems, which are difficult to construct. The only wind system demonstrated was a factory-built 4-KW system that had an installed cost of $3,897 and met only about one-fourth of the farm family's electric requirements. Its main value was demonstrating a direct tie-in with the rural electric system that serves Cedar County farmers (Aiken, 1979).

High cost was also a factor in a cooperator's decision not to build a methane digester that had been designed for his farm (Demmel, 1979). Cost estimates ranged from $8,000 for an 8,000-gallon digester (designed to produce 850 cubic feet of methane daily) to $17,000 for a 16,000-gallon system (designed to produce 1.700 cubic feet daily and compress it for use as a motor fuel). Other cooperators agreed that systems of this kind are not appropriate for the small crop-livestock farms they operate.

Another constraint was a lack of proven technology for construction of on-farm alcohol fuel systems. This precluded the construction of a farm-produced liquid fuel component that could have demonstrated the appropriateness of this technology on small farms. (The project received a small grant late in the second year from the U.S. Department of Energy to develop alcohol systems on three cooperator farms.)

A related constraint was lack of time for construction of projects. All of the cooperators had livestock operations that required several hours of labor daily. Experience showed that the best construction period was in late July and early August. Cooperators were also willing to attend daytime workshops in the winter and to construct projects when weather permitted.

Energy Consumption Patterns

The breakdown of energy use reported in this project differs from what is usually reported by agricultural researchers, partly because the farms involved do not have separate electric meters for household and farm use. Tractors and pickup trucks also were filled from the same gasoline storage tanks and some propane tanks provided fuel for both home and farm-related needs.

The overall energy use on the 48 farms in 1979, calculated on a gross energy basis, was as follows: 44% for motor fuels, 30% for electricity, 14% for space heating, and 12% for fertilizer.[2] These totals are based on data from farm energy record books. About 42% of all the energy purchased was for household use, although data on this point were insufficient.

It is difficult to compare this information with data categories from the national 1974 data base (Federal Energy Administration, 1976). This national study showed that fertilizer accounted for 31% of the energy used in agricultural production, irrigation for 13%, farm vehicles for 13%, preharvest field operations for 12%, and harvest operations for 10%. These figures, of course, do not reflect energy used for household purposes.

The energy figure for fertilizer reported by both the cooperator and control group farmers was low because large amounts of livestock manure were available and because the figure represents 12% of all energy used (not farm use alone). These farmers also included legumes in their crop rotations, which helped meet nitrogen requirements. It was

---

[2] Pesticide energy was not calculated. It is quite small on Corn Belt farms compared to fertilizer use. Electricity was computed at 10,240 BTU/kwh, or a conversion efficiency of 33%.

unusually difficult, as a result, for the cooperators to make a substantial reduction in energy use in the fertilizer category.

The greatest potential for savings was in motor fuels. But energy use reductions were limited mainly to minimum tillage, keeping tractors tuned and plows and discs sharpened, reducing evaporation by painting fuel storage tanks white and installing pressurized caps, and making fewer trips to town. It is interesting to note that cutting purchased motor fuels by one-third, which perhaps could be achieved by a combination of vigorous conservation and on-farm alcohol production, would save more energy than the total used for either space heating or fertilizer.

Comparative Energy Use Results

The main result of the three-year comparison is that the cooperators spent an average of $5,750 per farm for energy in 1979, which was 17% less than the $6,888 average for the control group farmers, for an annual savings of $1,138 per farm.

During the three-year period, the control group increased its energy use by 25% (Table 2), with motor fuels increasing the most in absolute amount, but with fertilizer increasing at the fastest relative rate (39%). However, because this was a period of sharply rising energy prices, energy expenses rose 63% (Table 3). Diesel fuel was the leading contributor both in absolute amount and relative increase, but all categories showed sharp increases.

In contrast, the cooperator farms had an energy use increase of only 9% and there was even a slight decline following the first year's increase of 11% (Table 4). The increase was well below that of the control group in all categories, and energy use for space heating actually declined by 16%. But despite the very modest increase in energy use, the energy bills for the cooperator farms went up by 44% because of the increase in energy prices (Table 5). Still, this was well below the 63% increase for the control group. As with the control group, expenses for diesel fuel increased most rapidly (128%), while propane costs declined because the space heating demand was reduced through insulation, wood burning, and solar heating systems.

The difference between the two groups' use of fertilizer was quite important. The 18% increase reported by the cooperators over three years compares with a 39% increase for control group farmers. This probably reflects the results of

Table 2. Energy Use on Control Group Farms[a]

Use Category	1976	1977	1978	1979	% Change (1976-9)
Electricity	268	280	322	348	+29.9
Space Heating	157	131	151	181	+15.3
Motor Fuels	438	522	545	529	+20.8
Fertilizer	115	123	167	160	+39.1
Totals	978	1,056	1,185	1,218	+24.5

[a] Average per farm in millions of BTUs

Table 3. Average Control Group Energy Costs

Use Category	1976	1977	1978	1979	% Change (1976-9)
Electricity	$ 788	$ 830	$1,027	$1,175	+49.1
Fuel Oil	191	178	189	322	+68.6
Propane	336	249	357	439	+30.7
Diesel	318	421	515	893	+180.8
Tractor Gas	792	903	1,051	1,316	+66.2
Car Gas	592	903	789	1,069	+80.6
Fertilizer[a]	1,220	1,252	1,272	1,674	+37.2
Totals	$4,237	$4,736	$5,200	$6,888	+62.6

[a] Cost of the fertilizer itself, not of the energy to manufacture it.

Table 4. Energy Use on Cooperator Farms[a]

Use Category	1976	1977	1978	1979	% Change (1976-79)
Electricity	313	325	318	346	+10.5
Space Heating	155	156	128	131	-15.5
Motor Fuels	408	495	488	466	+14.2
Fertilizer	100	103	107	118	+18.0
Totals	976	1,079	1,041	1,061	+8.7

[a] Average per farm in millions of BTUs.

Table 5. Average Cooperator Group Energy Costs

Use Category	1976	1977	1978	1979	% Change (1976-9)
Electricity	$ 944	$ 992	$1,083	$1,157	+22.6
Fuel Oil	209	262	231	343	+64.1
Propane	265	197	201	200	-24.5
Diesel	353	489	558	804	+127.8
Tractor Gas	617	789	807	928	+50.4
Car Gas	587	792	720	1,035	+76.3
Fertilizer[a]	1,011	1,045	882	1,283	+26.9
Totals	$3,986	$4,566	$4,482	$5,750	+44.3

[a] Cost of the fertilizer itself, not of the energy to manufacture it.

Table 6. Average Purchase of Energy for Cooperators

Use Category	1976	1977	1978	1979	% Change (1976-9)
Electricity (kwh)	31,076	32,248	31,429	33,291	+7.1
Fuel Oil (gallons)	544	611	468	529	-2.7
Propane (gallons)	686	583	522	462	-32.7
Diesel (gallons)	920	1,104	1,238	1,179	+28.1
Tractor Gas (gallons)	1,137	1,359	1,326	1,121	-1.4
Car Gas (gallons)	1,083	1,340	1,178	1,253	+15.7
Fertilizer (MBTUs)	100	103	107	118	+18.0

Table 7. Average Energy Purchases by Control Group Farmers

Use Category	1976	1977	1978	1979	% Change (1976-9)
Electricity (kwh)	25,943	26,990	31,333	33,809	+30.3
Fuel Oil (gallons)	496	414	413	497	+0.2
Propane (gallons)	871	734	898	1,013	+16.3
Diesel (gallons)	829	951	1,146	1,309	+57.9
Tractor Gas (gallons)	1,460	1,554	1,776	1,590	+8.9
Car Gas (gallons)	1,092	1,528	1,279	1,294	+18.5
Fertilizer (MBTUs)	115	123	167	160	+39.1

soil testing assistance provided to the cooperators in 1978, continued emphasis on nitrogen-fixing legume crops and efficient manure handling, and the adoption of manure composting.

Tables 6 and 7 provide a more detailed year-by-year breakdown of energy purchases by farmers in the two groups. The cooperators reduced their energy consumption in three categories (fuel oil, propane, and tractor gasoline) while the control group farmers increased consumption in every category. The increases reported by the cooperators were also less in all instances where increases were reported by both groups of farmers (electricity, diesel, car gasoline, and fertilizer).

Farm size remained almost constant for both groups during the three-year period (up to 364 acres from 357 for the cooperators and up to 361 acres from 357 for the control group farmers). Average annual inventories of grain, hay, silage, and livestock, which were calculated from record book entries by both groups of farmers, were comparable throughout the project period.

A breakdown of energy costs by different types of farming operations suggests that operators of diversified farms would probably find it easier to become energy self-sufficient than would operators of specialized dairy or hog farms. Table 8 shows, for example, that the net income returned for each dollar spent for energy was greater for diversified farms for all three project years. The amount spent for energy by operators of diversified farms also was considerably less for all three years.

Inability to insulate control group farmers from project information disseminated in the community and throughout the region turned out to be a serious design flaw. The high visibility achieved by the project in Cedar County had not been anticipated.

A 1978 project evaluation concluded that keeping detailed energy records and being exposed to word-of-mouth and published reports of project activities in the community made members of the control group much more aware of their energy costs and consumption patterns (Associates in Rural Development,Inc., 1979). The evaluators reported that 23 of the 24 control group farmers said they had become more aware of their energy costs since they had become involved in the project. The evaluation report also noted that ten farmers reported that they were planning to begin new energy projects in the near future and six had actually begun new energy-saving efforts.

Table 8. Average Energy Costs and Returns by Farm Type [a]

Farm Type	Energy Use in MBTUs	Energy Expenses Per Farm	Gross Income Per Farm	Net Income Per Farm	Net Income Per Energy Dollar
1977					
Diversified	905	$ 3,898	$ 28,036	$ 3,497	.90
Dairy	1,339	6,371	42,301	3,532	.55
Hog	1,080	4,732	31,145	3,186	.67
1978					
Diversified	965	4,228	38,320	7,833	1.85
Dairy	1,304	5,913	50,142	8,896	1.50
Hog	1,069	4,903	36,026	6,414	1.31
1979					
Diversified	974	5,215	45,903	10,175	1.95
Dairy	1,257	6,709	55,973	8,881	1.32
Hog	1,106	6,085	42,098	7,668	1.26

[a] This represents 20 participating farms (7 diversified, 6 dairy and 7 hog) for which complete income tax returns for 1977, 1978, and 1979 were made available. Both control group and cooperator farms are included.

This suggests that the energy consumption difference between the two groups would have been considerably greater if the control group could have been insulated from community and other outside influences. It also suggests that energy recordkeeping and community energy information programs may be sufficient to motivate farmers to take action to reduce energy consumption and adopt new energy-saving practices.

Individual Performance

The difference in energy consumption between the two groups of farmers would also have been greater if all the cooperators had built the full range of proven energy systems that appeared to be appropriate for their farms. The cooperators were encouraged to build every system that appeared to be appropriate for their farms and had unlimited access to cost share and technical assistance. The level of participation was an individual decision, however, and no attempt was made to push them into taking action they were reluctant to take.

It is clear that the number of energy systems built and the amount of energy saved would have been much greater if a "model farm" strategy had been employed. The self-selection approach used, however, made each cooperator responsible for determining the appropriateness of each innovation or practice adopted, and meant that the farmer had more of a commitment to making the innovation succeed. The wide range of energy systems constructed on cooperator farms, as a result, were all being used when the project ended.

All of the cooperators participated in project workshops and other educational activities, filled out energy record books, completed weatherization and soil sampling programs, and adopted other energy conservation practices. Nearly all completed one or more cost-shared devices or practices from a list of "minor innovations." The list included solar food dryers, windowbox solar collectors, wood heating, and minimum tillage.

Only ten completed one or more projects from a list of "major innovations." These included solar grain dryers, solar greenhouses, solar water heaters, solar wall collectors, wind generators, roof-mounted collector systems, and portable solar collectors. With the exception of the wind system, which was purchased, all of these innovations were built by the cooperators.

A staff analysis of economical innovations that could have been adopted on the 24 farms indicates that they would have reduced annual overall energy consumption by 917 million BTUs and cut annual expenditures for energy by $5,850. This averages out to $244 per farm. The estimated savings for the ten farms that adopted solar and other "major" innovations actually averaged about $120. These savings would increase each year, of course, as energy prices rise.

Table 9 shows that four of the most popular innovations would save enough money to pay for themselves in ten years and return an annual income benefit besides. Annual energy savings are based on figures obtained in monitoring cooperator-built systems.

Conclusions

The project offered an opportunity to observe systematically what a group of fulltime farmers would do when provided with technical and other assistance in developing their own energy programs. It was an opportunity to see how they perceived the economic value of energy-related changes, what factors they considered in determining the appropriateness of energy innovations and practices, and how they decided whether to invest scarce resources in energy systems

Table 9. Life Cycle Cost Analysis (10 Years)

Energy Innovation	Cost[(a)]	Annual Energy(b) Savings	Annual Dollar(c) Savings	Annual Income (d) Benefits
Solar Grain Dryer	$ 600	40	$ 319	$ 259
Solar Wall Collector	359	8	136	100
Portable Collector	1,740	24	191	17
Solar Water Heater	1,597	10	175	15

(a) Cost includes the 1978 price of materials plus a 10% loan amortized over three years.

(b) Annual output in millions of BTUs is based on actual records kept by farmer-owners.

(c) Average annual savings over a ten-year period assuming 10% inflation plus an annual price escalation of 26% for electricity and 5.7% for propane.

(d) The difference between average annual costs and average annual dollar savings.

that appeared economically feasible over a period of years but did not provide immediate economic benefits.

Although the amount of energy saved would undoubtedly have been much greater if a "model farm" strategy had been adopted, a project of this kind would have demonstrated little beyond whether a particular innovation or practice performed well and what it would cost under ideal conditions. This information was already available, in most cases, when the project got underway. The real need was to find out more about the adaptability of low-cost energy systems and the way energy decisions are made by farmers.

Farmers with more available capital probably would have done some things differently. But they, too, would have had to make decisions about appropriateness and economic feasibility and the factors they would have considered would have been similar.

The project does not begin to answer all the questions that will arise as farmers move toward energy self-sufficiency. But it does identify many of the practical problems that will be encountered and provides some useful information that will help show the way.

References

Aiken, Rob. 1979. Wind electric power generation and connection to a rural electric cooperative in northeast Nebraska. Paper presented at a USDA-DOE workshop on wind energy applications in agriculture. Iowa State University, Ames, Iowa.

Associates in Rural Development, Inc. 1979. Evaluation of the Small Farm Energy Project at the Center for Rural Affairs. Burlington, Vermont.

Blobaum, Roger. 1980. Small is renewable: the impact of energy policy on farm structure. In Perspectives on the Structure of American Agriculture. Rural America, Washington, D.C.

Buttel, Frederick H., William Lockeretz, Martin Strange, and Elinor Terhune. 1980. Energy and Small Farms: A Review of Existing Literature and Suggestions Concerning Future Research. National Rural Center, Washington, D.C.

Demmel, Dennis. 1979. Anerobic digestion for methane generation on small farms in northeast Nebraska. Paper presented at the Mid-America Bio-Mass Energy Workshop. Purdue University, Lafayette, Indiana.

DOE. 1978. Solar Energy for Agricultural and Industrial Process Heat. Report No. CS-0053. Washington, D.C.

Federal Energy Administration. 1976. Energy and U.S. Agriculture: 1974 Data Base. FEA/D-76/459.

USDA. 1979. Energy alternatives and actions for U.S. Agriculture. Report Prepared for the National Agricultural Research and Extension Users Advisory Board, Washington, D.C.

USDA. 1980. Economic Indicators of the Farm Sector: Income and Balance Sheet Statistics, 1979. Statistical Bulletin No. 650. Washington, D.C.

Williams, Jim. 1979. Statement to Subcommittee on Agricultural Research and General Legislation, Senate Committee on Agriculture, Nutrition, and Forestry. Washington, D.C.

Folke Dovring

4. Effects of an Alcohol Fuels Program on Farm Economics

The effects that an alcohol fuels program based on biomass will have on farm economics will vary with the scope of the program, the scale of operation prevailing in it, and the choices of feedstocks, process fuels, and products. Likely repercussions on farming as an industry, and on the economics of individual farms, will range from a price support effect, to economic diversification, to changes in land use, cropping systems and soil conservation. To what extent biomass fuels programs may be economically justified is a separate topic that will not be explored here; the answers will depend on assumptions about opportunity costs (present and future), which are not the same for all situations (Dovring, 1980), and on pending technological findings about the efficiency of various fuels used in various ways. The general problem of resource tradeoffs between solar and mined sources of energy has been treated elsewhere (Dovring, 1981). The following assumes that some alcohol fuels program will be adopted, for whatever reasons.

The scope of the program could be anywhere from insignificant to substantial, and the consequent displacements would vary accordingly. Scale of operation could favor large industry, small industry, or farm-level production. Each has its specific advantages and disadvantages, which include not only the conventional economies and diseconomies of scale and size, but also differences in the prices assigned to inputs and the values assigned to outputs when inputs and outputs both involve transportation. Further differences relate to the feasibility of using bulky feedstocks and local fuels.

Most of the effects are likely to be different for different feedstock types. We will therefore begin with these, and then follow up by discussing types of effects.

Feedstock Types

Considerable discussion about plausible feedstocks for biomass alcohol fuel (David et al., 1978; Doering and Tyner, 1980; Herman, 1981; Meekhof, Gill and Tyner, 1980; Office of Technology Assessment, 1980; Solar Energy Research Institute, 1980; Tyner, 1981; United States Department of Agriculture, 1980; DOE, 1979) has led to the general view that for agriculture there are three main types of feedstock: grain crops for ethanol, silage crops for ethanol, and wood for methanol. The alternatives will be discussed in sequence, before generalizing the conclusions.

Grain Crops for Ethanol

Potential production and effects on crop prices. This is the type of feedstock already in use or being considered in most plants now operating or planned. So far, such production is small on a national scale. In fact, it is smaller than might be deduced from the "gasohol" supply on the market, since some of the alcohol has been imported from Brazil. For late 1980, fuel alcohol distillery capacity in the United States was given as 120 million gallons a year (Herman, 1981).

For a given size program, grain crops would have the greatest impact on the farm sector of any type of feedstock. The impact would be further affected by the size and location of the distilleries. Initiatives have been taken to build distilleries of many different sizes and with different ties to other economic activities, such as wet milling, cattle feeding, etc. If current investment plans materialize, within two or three years we may have enough alcohol production from grain to make an impact on grain market prices and feed markets, and indirectly on land use.

There is no one volume of alcohol from grain that should be analyzed. It is clear that alcohol from grain could never become more than a minor component in the supply of liquid motor fuel in the United States. An objective sometimes discussed is to replace one-tenth of all gasoline with ethyl alcohol. If this were all to come from grain, it would require about half the current annual harvest of feed grains in this country. The resulting feed by-product (about 30% of the grain, by weight) would exceed all the soybean meal presently being produced in the country. The soybean industry would be virtually wiped out and vegetable fats would have to come from other crops (corn, sunflower, peanuts). Because soybeans would be displaced by corn and sorghum, energy intensity of field crop production would be substantially increased through much larger application of nitrogen fertilizers, as will be discussed further in Chapter 5. Nowhere near all of the feed by-product could be used as high-protein feed, and some of it would be used as energy feed, which has lower value per unit of weight (Berger, 1980). From these indications alone, it is clear that the 10% objective could not be achieved entirely, or even mainly, from grain crops. Much lower targets have also been discussed, mainly from the viewpoint of the impact they might have on the price of food.

One such target is 4 billion gallons a year, as discussed in a report from Schnittker Associates (1980). The feedstock would be about 1.6 billion bushels, or 40 million metric tons, of grain, or about 1/6 of the annual feed grain production in the United States in recent years. The feed by-product, about 12 million metric tons, would still displace a large part of the soybean meal currently produced in the country (Wisner and Gidel, 1977). The report estimates that this volume of alcohol production would not affect food prices much--they would rise merely 5 to 6%. Whatever the validity of this estimate, concerning which there are several opinions, it should be pointed out that such a change in food prices would include a several times larger increase in the prices of feed grains. Hence, such a volume of distillery activity would not only represent a large demand for feed grains and a substantial dislocation in the feed markets, but would also act as a price support for grains, and indirectly for field crops in general. Such price supports would have quite different meaning for large and small distilleries.

Types of facilities. In addition, their effect would differ strongly for different types of operations. Some large food processing firms, such as Archer Daniels Midland of Decatur, Illinois, engage in complex "corn processing". The oil and the protein are removed first and used in the production of highly processed foods. The starch is converted to sugar, fermented, and then distilled for alcohol, leaving some feed by-product. Secondary by-products, such as carbon dioxide and fusel oil (consisting of higher alcohols such as butyl and amyl), are also recovered and put to economic use. To such an operation, the price of corn is important, of course, but not decisive, since some of the joint products are foods whose prices must also rise with a rise in the price of corn. The situation is different for distilleries which aim foremost at producing alcohol fuel and for which the high-protein livestock feed is a by-product. Some oil companies plan large distilleries (20-60 million gallons per year) primarily to obtain an octane booster for their gasoline; other independent distilleries with a capacity of a few million gallons per year also aim essentially at the gasohol market. For such distilleries a rise in the price of grain because of increased distillery use of grains would be a negative impact of their own success. To explain how this is different from the case of small, farm-based distilleries, some of the economic differences between the latter and the industrial plants must be discussed.

Economic comparison of small and large facilities. There are few empirical studies of operating alcohol distilleries. At the University of Illinois, a study was completed recently which compares the technical and economic results of three prototype plants currently in operation (Dovring et al., 1980). The plants analyzed include a farm-level plant of 10 to 15 thousand gallons a year annual capacity;1/ an intermediate-size plant, indicated for a farmer consortium or a rural cooperative, with an annual capacity of ¼ million gallons (and capable of modular expansion to ½, 3/4, or 1 million gallons); and a small industrial distillery of 2½ million gallons annual capacity, capable of modular expansion to 10 million. The latter uses a new distillation technique, the ACR process, which saves energy by distilling a mixture of ethyl alcohol, water, and gasoline (Chambers, 1979). In contrast to conventional distilleries that produce pure ethyl alcohol, this product is suitable only for further mixing with gasoline to produce gasohol.

In comparing these three plants, some significant differences emerged. The larger plants have better energy efficiency, but how much of the ACR process efficiency was due to size and how much to the process has not been determined.2/ There are also increasing returns to size in the use of capital. Use of labor is complicated by the difficulty of placing a price on farmers' self-employed labor, especially in agriculture's off seasons. Offsetting the larger plants' advantages in use of capital and energy, the smaller plants have advantages in the dollar balances because they save on the

---

1/ In the central Corn Belt, it takes about 10 gallons of diesel fuel to cultivate and harvest an acre in row crops, or 15 gallons of ethyl alcohol. Thus, 10 to 15 thousand gallons would serve a farm of 650 to 1000 acres, or a group of smaller farms with the same aggregate acreage.

2/ Another report on a farm-level distillery (in Colorado) indicates fuel efficiency higher than on those in Illinois, and not far below that of the ACR process (Jantzen and McKinnon, 1980). However, the technique by which the energy balance was established at the Colorado plant may not be fully comparable with that in the Illinois study, and so the validity of differences could not be verified.

trading margins for the grain used as a raw material and the by-product feeds. A farmer-owned or farm consortium plant need only charge the farm gate price of grain (the income foregone by not selling the grain), while an industrial plant of whatever size must pay the wholesale market price, which is generally about ten percent higher. Credits for by-product feeds are also quite different. Large plants must sell at wholesale prices, while farmers or farmer groups who utilize the feed on their own farms can credit themselves with the retail price of such feeds, which is generally one-third higher than that of the wholesale market.3/ Because of all this, the value of the feed by-product on small distilleries sometimes exceeds that of the alcohol (it is more than half of the value of all outputs), while at large industrial plants, the value of feed by-product comes to only about half of the value of alcohol (one-third of the value of all outputs). The dollar balance on the small plants comes to about the same as on the large one: the price advantages balance the lower efficiency in use of capital and process energy.

The small, farm-level or farm-consortium distillery is also far less sensitive to the price of grain than is true of the large distilleries. In the latter, the price-increasing effect of a substantial volume of distillery activity reduces the margin of profit. At the farm level, such an effect is of much less significance. In fact, a high price of grain is the grain farmers' main source of profit to begin with, and the price-increasing effect may add more profit to the farm's main enterprise than it takes from the auxiliary enterprise, alcohol distillation. This point is important enough so that many farm-level distilleries will feel free to vary their distillery activity in inverse proportion with the price of grain, a market reaction which the large industrial plants are much less disposed to follow because of their larger overheads in capital and skilled labor. However, this advantage is limited by the fact that not using an on-farm system imposes considerable fixed costs. Also, the farmer must be

---

3/ A further complication in evaluating by-product feeds comes from the fact that on-farm plants would probably retain only the solid portion of the feed (DDG, distillers dried grains without solubles). In contrast, industrial plants can dry the by-products, thereby retaining both the solids and solubles (DDGS, distillers dried grains with solubles). These two forms have different value per unit weight, since drying the solubles lowers the feed value for ruminants (beef and dairy cattle). On the other hand, the solubles are particularly valuable for non-ruminants, such as hogs (Berger, 1980).

prepared to switch his livestock operations away from by-product feeds and must be able to return his equipment to using fossil fuel if he chooses to stop producing alcohol for an extended period, unless he has stored up a large supply of alcohol.

The last point also leads to another conclusion: that the market effect of the large and the small distilleries is likely to be different in yet one more respect. It has been said, as a negative reflection on the grain alcohol enterprise, that it would tend to subtract from the domestic and international grain markets a more or less constant amount of grain, leaving the normal markets to absorb the entire impact of year-to-year supply variations due to weather. This conclusion appears quite logical with regard to the large industrial distilleries, which will tend to maintain full capacity utilization, as already mentioned. The farm-level plants, with their close ties to the total farm business, are more likely to be flexible regarding their volume of activity. Alcohol can be kept in storage indefinitely--more so than grain.

The monetary costs on the large and small plants were not found to be significantly different. They have different economic meaning, however, because the products do not go to the same market. The large plants aim at the gasohol market, on which their output is subsidized by tax relief (Federal, and additional relief in many states). The small plants generally can not reach the gasohol market because their output is not 100% (200 proof) alcohol, but some lower grade, since removing the last few percentage points of water is very difficult in small plants. Selling to other distilleries would impose a price discount. The rationale of the farm-level or farm-consortium distillery is instead to supply fuel for use right on the farm. In modified diesel engines, 180 proof (90%) alcohol is said to be preferable because the steam pressure is said to enhance fuel efficiency. The farmers' motive in owning their own distilleries is in more security of fuel supply rather than in earning increased net income. The easiest ways of using alcohol in current farm machinery permit only a partial replacement of gasoline or diesel fuel. Substantial fuel self-sufficiency therefore will require some investment to either replace or modify existing machinery.

Energy comparison of small and large facilities. It remains to discuss the fuel balance of distilleries of different sizes. Overall, gasohol does not have a very favorable energy balance; there is gain in high quality (liquid motor) fuel, but the heat value of the output and the input energy are close to each other (Chambers et al., 1979). The larger distilleries could be expected to be more efficient if the various plants all used the same fuels. Larger tanks have a more favorable surface/volume ratio, and thereby lose less heat to the surrounding air. Also, certain heat recovery techniques are too expensive for smaller plants. But the energy advantage of larger plants is significant only with continued use of high-grade fossil fuels for process energy, such as propane, fuel oil, or natural gas. A switch to coal would make the difference in process fuel efficiency less interesting to national economic policy, but small farm-level or farm-consortium level distilleries could also use locally-available fuels that have much lower opportunity costs.

Such fuels could include scrub wood, or corn and soybean stover, which are too bulky to transport long distances (for fuel use, in any event) and are difficult to store in large quantities. However, use of these fuels on a small scale either is extremely labor-intensive or imposes considerable capital costs. Another plausible local fuel is methane from anaerobic digesters using manure from the same or nearby farms--including manure derived from feeding the feed by-product from the same distillery that is using the methane as process fuel. Such use of local fuels would greatly improve fuel alcohol's energy balance with regard to scarce, high-grade energy sources.

Effects on location of cattle feeding. A potential effect of using grain as the alcohol raw material is the possibility of attracting back to the Corn Belt some of the cattle feeding activity that in recent decades has moved to the southwestern states (western Kansas to southern California), but which now is experiencing difficulties from rising freight rates for feed grain from the Corn Belt and also from the scarcity of irrigation water and higher pumping costs. When added to these problems, the lessened supply of Midwestern grain and larger supply of wet distillery residue on farms in the Corn Belt could help divert some of the livestock feeding to grain areas with on-farm distilleries. Large industrial distilleries will have less of this effect on the location of animal production, since their dried by-product feeds could be shipped to cattle in other regions.

The recent opinion in the literature generally is that grain alcohol may serve as an opening wedge but that grain crops are likely to remain a minor feature even in the alcohol industry, and an even smaller one in the energy supply of the United States. The various materials would have different effects on soil erosion and foreign markets, as will be discussed later.

Silage Crops for Ethanol

This type of production is as yet essentially untested, but there is little doubt that it can be viable under the right operational conditions. The technique of ensilage is of course well known and it appears to be generally agreed that losses in storage can be kept moderate. Effects on farm economics and on soil conservation can be quite different from those of grain crops.

The botanical inventory of potential silage crops is large. In addition to corn and sorghum, both of which are already grown to a substantial degree as silage feed crops, silage for ethanol could be obtained from a large number of perennial species. These include tall forage grasses, alfalfa, and Jerusalem artichoke (the "Kansas sunflower", *Helianthus tuberosus*). When harvested at the right season, silage contains mainly starch, not much cellulose and no lignin, and not much protein either. If species are chosen that are best for each area, soil type, and micro climate, silage crops should be able to yield more starch and hence more alcohol per acre than grain crops. They would also require less expense for fertilizers, and less tillage in the case of perennial species. Such crops can often be grown successfully on land less fertile than first-rate grain land, and on land vulnerable to erosion. In the latter case, the preferred species are perennials that develop extensive root systems to stabilize the soil.

The total potential for silage as feedstock is now estimated to be larger than from grain, in the range of 1 to 4 quads per year, which corresponds to 0.2 to 0.7 billion barrels of oil equivalent (Tyner, 1981). The range depends in part on assumptions regarding competition with other land uses, which to some extent would be with cultivated crops but mainly would be with tree production. Even the lower bound of the estimated range could supply most, if not all, of the motor fuel needed in U.S. agriculture.

Silage crops would not be suitable for large distilleries because their bulk makes transportation expensive, and also because management of ensilage is best done in silos of moderate size. Thus this class of feedstocks is much more favorable for a small-scale operation, either on an individual farm or a farmer consortium or cooperative.

This type of feedstock also may not necessitate a direct connection with animal husbandry. The distillation residue is likely to be lower in protein than in the case of grain.

The wet stillage, after filtration to reduce water content, could then be placed directly in anaerobic digesters to produce methane to be used as process fuel for the distillery. The residue from methane production would still have some fertilizer value. In this manner, farm-based small distilleries could have a very favorable fossil fuel balance. Instead, there would be a substantial land cost for the energy --the commitment of acreage specifically for energy production is larger in this case since the process would generate only fuel alcohol, and no by-product feeds. (The fertilizer residue from methane digestion counts as partial recycling, not new product).

This larger commitment of land would be partially compensated, however, by the fact that the land might otherwise be unused or used only in low-value ways. Also, there would be less impact on the markets for food and feed products. The use of low-grade land represents an expansion of the land resource by opening up such lands for more intensive uses without excessive erosion. If distillery residue is used directly for methane production, there will also be no impact on the markets for livestock feeds, and hence soybeans would be essentially unaffected. The supply of grain for both domestic and export markets would also be affected only to the extent that current arable cropland is used to grow silage. Such land use can still have some price-supporting effect in regard to grains and other cultivated crops, but without any direct disruption of the markets for those crops. There would be no increase in the variability of grain supply.

Other Crops for Ethanol

Several other crops besides grains and silage crops have been considered as possible alcohol raw materials, especially sugar cane, sweet sorghum, and sugar beets. Some attention has also been given to other root crops such as fodder beets and potatoes.[4/]

In the United States sugar cane is grown only in limited areas. Only Hawaii has the advantage of supplying feedstock

---

4/ Estimates of alcohol potential usually also include alcohol that may be produced from waste materials such as citrus rinds, cheese whey, and municipal solid wastes. Such feedstocks will contribute only marginal quantities to the country's fuel supply and their impact on the agricultural economy can be regarded as negligible.

all year long, but Hawaii has no spare land, and if anything the cane areas are contracting. The two main cane areas on the mainland are in Louisiana and Florida, both of which have limited seasonal operation, which would be disadvantageous for alcohol distilleries. If sucrose from cane were to be supplemented by other (starch) feedstocks to fill out the work year, the simplicity of fermenting only sucrose would be lost. Also, equipment for saccharification of the starch would be needed, but again only for operation in some seasons. Use of the crop residue (the "bagasse") as process fuel in the distilleries, as in cane sugar mills, is an advantage, but again one that is only available during the cane processing season. Distilleries using sugar cane appear suited for relatively large scale of operation, for the same reasons as in cane sugar production. Any impact on the farm sector as a whole would be limited, since the small sugar cane areas are not likely to expand much. Since the U.S. currently imports close to half its sugar, production of alcohol from sugar cane without a corresponding increase in cane production (or a reduction in sugar consumption) could mean dependence on foreign sources of sugar, unless corn sweetener production is stepped up further (it is already increasing rapidly).

Sweet sorghum has been investigated recently and in some analyses is envisioned as having a substantial alcohol potential (DOE, 1979). On the positive side, sweet sorghum can be grown in large parts of the United States. It would compete with other field crops but would not generate additional feed supplies. As with sugar cane, the crop residue would be burned as process fuel. Since the feedstock would be sucrose and not starch, long-term storage appears unlikely. Hence distillery operation would have seasonal constraints similar to those with sugar cane. This might point to small-scale, farm-based distilleries as the likely solution, although small-scale facilities to process sweet sorghum have technical problems. Because high grade farmland would be used, rather than the marginal lands suitable for silage crops, it now appears less likely that sweet sorghum would be a major feedstock for fuel alcohol.

Sugar beets for sucrose appear less likely than beets and other root crops for starch. Possibly such crops could also be stored as silage. As with sweet sorghum, the disadvantage is that these crops would occupy land drawn entirely from the supply of prime cropland. Size of operation would likely be small, for the same reasons as with green silage crops, and the purpose of such production would also be mainly to supply fuel to operate the farm.

Crop Residue As Alcohol Feedstock or Process Fuel

Much attention has been paid to the possibility of utilizing crop residues such as corn stover, corn cobs, soybean stover, and wheat straw for energy production. The limits on how much residue can be removed from fields without causing destructive soil deterioration are discussed in Chapter 6. Equally important limitations are the residues' bulkiness and deterioration in storage. Whether as feedstock or as process fuel, crop residues could only be used in plants of small or moderate size.

Stover is mainly cellulose, and for ethanol production it would have to be subjected to a much more complex series of chemical transformations than those needed for grains or green silage (Ladisch, Ladisch, and Tsao, 1978). The cost of ethanol production from cellulose is still much higher than with starch (Miller, 1980; Spano et al., 1980). Alcohol yield has also been quite low in experiments done so far. Stover as a feedstock appears as an unlikely alternative for farm-level production because of the sophisticated technology involved.

Use of crop residues as process fuel is closer to being realistic, at least for a small plant drawing on nearby sources. One small farm-level distillery in Iowa reported using corn cobs as boiler fuel. This is done most conveniently if the corn is harvested on the ear, rather than with a combine, and also stored on the ear in an oldfashioned corn crib. It is then shelled as grain is required for fermentation. This will produce corn cobs at the rate needed to supply heat for distillation, and the cobs would be well stored and dry. Stover is often so moist that a substantial part of its heat value is used up to dry the fuel itself. The possibility may also exist of blending crop residues with manure or distillery residue for digestion into methane, whether for process fuel in the distillery or for other purposes.

If kept within limits, withdrawal of crop residue would cause no inconvenience other than the need for additional fertilizers. When burned in a boiler, only the ashes would be left behind, while blending with other materials for digestion into methane would leave behind more residue that could be used as fertilizer.

Wood for Methanol

Of all feedstocks for biomass fuel, wood has by far the largest potential (Tyner, 1981; DOE, 1979). Over the medium term (say, until year 2000) it is estimated that methanol from wood could cover about half of the current use of liquid motor fuel in the United States (DOE, 1979). It is also the source that has the least potential impact on agriculture for such a large volume of production.

Wood can of course also be converted to ethanol, by some cellulose-to-starch process (Ladisch, Ladisch, and Tsao, 1978; Miller, 1980; Spano et al., 1980). Recent findings indicate that gasification followed by methanol synthesis is more advantageous and yields more liquid fuel per unit of feedstock than conversion to ethanol (Hannon and Blanco, 1979). Solid residue is utilized as part of the process fuel in the conversion, and thus there are no by-products leaving the plant except ashes (which can be used as fertilizers).

Methanol production from wood delivers more energy per unit area of land than any of the alternatives. Trees fix more energy per unit area per year than do annual crops under comparable conditions (Lieth, 1972). Trees also can produce well on land that is not particularly valuable for field crops. They can draw on deeper soil strata than do most annual crops, and at least in the near term trees do not require fertilizers to the extent that annual crops do. Continuous, intensive tree production may modify that. Costs and energy use for planting and harvesting trees are also less than with annual field crops, so that the total energy requirement for wood production is less than for other biomass feedstocks.

In regard to soil conservation, trees offer some of the same advantages as perennial silage crops. Most estimates of potential for wood and for silage assume large overlaps between these two land uses. Either may have some advantage over the other on some kinds of land. For arable land temporarily used for energy feedstock production, silage crops may be the better choice. Many areas not suitable for cropping will produce more feedstock when planted with trees. "Tree farms" have been proposed, and some will no doubt be planted. In many farming areas of the United States, land suitable to produce wood as an alcohol feedstock includes more than the recorded areas of farm forest, which of course often can be upgraded by planting more productive species and by intensive management. Much of the "other land" in American farms can also be utilized in this way. In many cases, tree farming would represent increased intensity of land use. A general issue in using wood for alcohol is that much wood is worth more in its traditional uses. However, this does not apply to certain kinds of production, such as short-rotation tree farms, which produce wood that is not suitable for most non-alcohol uses.

Another plausible form of on-farm wood production may be shelterbelts along the edges of fields, oriented to catch the prevailing wind in each area. Shelterbelts need only provide a modest increase in crops yields for the wood to be produced virtually without land cost. Selective felling and replanting (or coppicing) can maintain the shelterbelt while yielding a sustained supply of wood.

Wood production on farms would represent a measure of economic diversification in many farming areas, which would be particularly useful in the areas of grain monoculture such as the central Corn Belt and parts of the Great Plains. Tree felling and timber transportation is winter work, and wood can be stored in the open, more conveniently than stover.

It is not clear at this time what size of plants would be most advantageous. Compared to ethanol from most crops, methanol from wood lends itself better to larger size installations. Whether plants can be small enough to be owned and operated by farmer consortia or cooperatives is worth exploring.

Types of Effects

The analysis by types of feedstock showed that the effect of an alcohol fuels program on the agricultural economy would depend partly on the raw material used. The various raw materials differ in the size and scale of operation, the choice of process fuel, and the transportation requirements of the material itself as well as of the process fuels and products. Areas that will be affected include: commodity price supports; land-use; production systems and volume of food production; soil conservation; and the diversity of the rural economy. Changed relations to the world food and agricultural economy is a separate topic pursued briefly at the end of this paper.

Price Support

An obvious consequence of withdrawing either food or food-producing land from the food and feed markets is upward pressure on the prices of the goods whose supply has been reduced. At current production levels, fuel alcohol has no measurable price-support effect. At what quantity such an effect might be felt is a matter of conflicting opinions.

This would depend on what feedstock is used. With grains, not only would there be less grain for the market but also more grain would be grown. This means higher real costs of producing grain, whether by making production more intensive (higher fertilization rates) or more extensive (growing grain on land with lower yield potential). This in turn means that the higher cost to the consumer is at least in part absorbed by higher cost to the farmers. Thus farmers receive less additional profit than if the whole increase in market price of grain were due to scarcity (without any change in cost of production). With silage and wood as raw materials, the cost of production is likely to be lower to the farmer, in part because the production comes from marginal land not suited to annual field crops. Any increase in grain production on land still in field crops would come at the intensive margin (fertilizers) which is likely to impose less of a real increase in production costs. Moreover, the silage and wood options are not likely to displace soybeans from the cropping system, thus avoiding the increase in the use of nitrogen fertilizer (and therefore energy) that would follow from such displacement.

The price support effect would have different consequences for large and small plants. For large industrial plants using grain, the price support effect would be a disadvantage--a penalty for success. For small plants owned and operated by individual farmers or by farmer consortia or cooperatives, the higher cost of alcohol feedstock would be more than compensated by the advantages to farmers of higher prices for their grain. Plants using silage as feedstock would almost certainly all be of small or moderate size, and would operate in close relation to the farms producing the feedstock. Methanol plants processing wood (or stover) would probably be relatively large, but any effect of higher crop prices would be, at most, indirect and weak.

Land-Use and Production Systems

Grains as feedstocks would lead to a large shift in the direction of cereals, especially corn and sorghum. In some areas, (Northern Plains, Pacific Northwest) barley and white wheat could also be expanded to serve as feedstock, while in the Plains bread wheat could be displaced onto somewhat less productive lands to make room for more corn and sorghum. The most important change would be that soybeans would be severely reduced, perhaps almost to the point of being wiped out, depending on the volume of grain alcohol production. To replace the supply of vegetable oil from soybeans, other oil-bearing crops would have to be expanded. More corn oil would require more of the kind of "corn processing" plants typified by Archer Daniels Midland, of Decatur, which is not what oil companies and others are planning for large distilleries. Other oil bearing crops such as sunflower also produce feed by-products, although in lower proportion than does soybean.

Thus the enlarged supply of feed by-product from distilleries would still meet competition, although less than that represented by soybean meal, so that a large grain alcohol industry would tend to depress the prices of concentrate feeds.

This is somewhat of a paradox, because in total, any large grain alcohol activity would mean higher food prices (Schnittker Associates, 1980), and a somewhat reduced total volume of food production. If the reduced total supply of crops available for food contains more concentrate livestock feeds than before, more of these feeds would have to be used below their value, that is, merely as sources of feed energy rather than of protein.

Soil Conservation

If more grain is used than is available because of occasional oversupply, this will lead to more intensive use of cropland, including marginal cropland and cropland vulnerable to erosion. There are, in fact, more row crops in the United States than is compatible with soil maintenance. For instance, in Illinois it was estimated that some 900,000 acres of cropland ought to be converted to grass or forest (Illinois Environmental Protection Agency, 1979, p. 7). Increased grain cropping, especially of corn and sorghum, would lead to more of this undesirable system.

Silage crops can also be chosen from among row-crop cereals, such as corn and sorghum. If this happens, the effects would be even worse, because silage removes the whole plant, including the stems and leaves, which are left behind when only the grain is harvested. But most, if not all, of the silage crops ought to be perennials sown on land which is marginal for arable crops. To this extent, silage crops will contribute to stabilizing erodible lands and reduce the flow of silt into streams and reservoirs. The same obviously goes for wood production under appropriate management. In both cases there will have to be increased attention to soil building by fertilizers and mulches to counteract possible soil exhaustion and to maintain productivity.

Diversification of the Rural Economy

Alcohol distilleries in rural areas will mean additional economic activity and income, more employment in off seasons, and risk avoidance by diversification. Diverting part of the crop potential (grain or silage) to alcohol production means an additional market, whatever the size or ownership of the distilleries. It could also mean an additional measure of flexibility, in cases where distillery activity (both from grain and silage) can be varied inversely with variations in crop prices.

Diversification should be of particular significance to specialized cash-grain areas, as in the central Corn Belt and in portions of the Great Plains. The current trend toward a "grain desert" of highly specialized and rather large farms in much of Illinois and parts of surrounding states means increased risk in farming, increased soil erosion, and lower land-use intensity because of non-use of former pasture lands. The trend toward larger and more capital intensive farms does not solve the basic problem of grain monoculture (Dovring and Yanagida, 1979). Symptomatic of the need for diversification is the occurrence of small manufacturing workshops on many cash-grain farms. Alcohol distilleries can represent another outlet for such activity. Because of the tie to the farm business--use of the alcohol as fuel in farm machinery and trucks--such distillery activity also represents less risk in farming because "marketing" of the output is secured in advance. Correspondingly, this reduces or removes the need to purchase diesel fuel, the price of which can only go up. A secure fuel supply is also a gain with this approach, but it must be traded off against the considerably higher cost of the alcohol.

All of these advantages will be more securely tied to the farms if the leading feedstock becomes silage rather than grain. Prevalence of large industrial distilleries using grain would withhold most of the diversification advantages from the rural areas.

A parallel question is the use of locally available fuels. Small grain distilleries may use corn cobs, stover, locally produced scrub wood, or some combination. The possibilities of producing methane gas from anaerobic digestion and using that as process fuel in distilleries is also an option mainly for small distilleries in rural areas. By lowering the cash cost for distillery operation, use of such fuels represents yet another element of diversification and risk avoidance.

Methanol from wood is likely to have a different set of effects on the rural economy. Since the total potential is much larger and a large part of the fuel would have to be sold to sectors other than agriculture, much of this activity eventually will be handled by industrial concerns that are less tied to farming areas than in the case of crop based ethanol. Even so, the bulkiness of the feedstock may lead to somewhat greater decentralization than what is envisioned by some oil companies in their plans for plants making ethanol from grain. In any event, growing trees would diversify the use of land, often generating more income than from present scrub forest and unused former pastureland. Felling and transportation of timber is winter work and could increase employment round the year, at the same time helping to diversify farmers' income.

International Perspective

The United States farm economy reaches across the world, with farm exports vastly exceeding farm imports. Specifically, the U.S. contributes about one-half of all grain that moves in international trade. A large biomass-based alcohol fuels program would be likely to cause some reduction in production of food and fibers, at least in relation to the volume of production that would be forthcoming without such an alcohol program. This is easily perceived when crops already produced are diverted to distilleries. But it also would follow, even though more indirectly and to somewhat lesser extent, from alcohol production from either herbaceous perennials or trees. Such land uses are likely to take over some marginal cropland and thus to affect ordinary crop production adversely.

The effect on international markets is likely to be different with the several alternatives discussed above. Grain alcohol from large industrial plants would probably increase the weather-related variability of exportable supplies. Farm-based grain distilleries are less likely to do this, while distilleries using silage crops or wood are likely to leave the variation in crop output unchanged or slightly reduced, since they would displace crop production on marginal lands that are more variable in their yields than are prime farmlands.

The economics of an alcohol fuels program is tied to an ethical question that is frequently raised--is it unethical to burn food? No more so than to burn (or to use as energy) something else produced where grain might have been raised. This brings to mind what happens with animal draft power, a system which is still important in many parts of the world. The total land resource can be used for a combination of purposes, and what share of it should go into fuel production will depend upon the tradeoffs involved, including impact on the farming economy (Dovring, 1980).

In its extreme form, the ethical question is posed with regard to foreign markets: What if biomass fuel production in the United States also means that the country will contribute less to feeding the world, and especially the world's low-income countries? Most of the food and feed exported from the United States goes to high-income countries (Europe, Japan) and for them, the basic answer is the same as for U.S.: there is considerable scope for modifying diets and using more food based directly on crops, rather than animal products. Such adjustments will occur to the extent the tradeoffs between food and energy--both of which are compelling necessities, up to a point--will force them. The low-income countries do not have as much of a choice in this direction, but for them we must ask the question whether food imports from rich countries is the best way of meeting their needs. Most of them have unused yield potentials in their own farmlands, and we might do them a greater favor by sending them fertilizers rather than food. This could be more useful to their continuing economic development. American farmers would then lose some part of their present role in "processing" chemicals (a primary input) into grain (an intermediate good on the way toward complete food production). But they might gain more than they would lose. Export agriculture is always plagued by extreme variability of markets. More diversified local economies would add up to a more diversified national rural economy. With more diversification and better soil conservation, America's farmers would gain more of the economic stability that federal farm policy has pursued for decades at considerable financial cost.

References

Berger, Larry L. 1980. Nutritional Value of Distillers Feeds for Livestock. 13 pp. Department of Animal Science, University of Illinois at Urbana-Champaign (mimeo).

Chambers, Robert S. 1979. The Small Fuel-Alcohol Distillery: General Description and Economic Feasibility Workbook. ACR Process Corporation, Urbana, Illinois.

Chambers, Robert S., R.A. Herendeen, J.J. Joyce and P.S. Penner. 1979. Gasohol: does it or doesn't it produce positive net energy? Science 206:789-795.

David, Milton L., Geneva S. Hammaker, Robert J. Buzenberg and John P. Wagner. 1978. Gasohol: Economic Feasibility Study, Final Report. SAN-1681-T7. U.S. Department of Energy, Washington, D.C. Available from National Technical Information Service.

DOE. 1979. The Report of the Alcohol Fuels Policy Review. DOE/PE-0012. U.S. Department of Energy, Washington, D.C.

Doering, Otto C. III and Wallace E. Tyner. 1980. Alcohol Production from Agricultural Products: An Update on the Facts and Issues. EC-511. Cooperative Extension Service, Purdue University, Lafayette, Indiana.

Dovring, Folke. 1980. Export or burn? American grain and the energy equations. Illinois Business Review 37(May):4.

Dovring, Folke. 1981. Energy production and natural resource tradeoffs. In W.E. Jeske (ed.), Economics, Ethics, Ecology: Roots of Productive Conservation. Soil Conservation Society of America, Ankeny, Iowa.

Dovring, Folke, Robert Herendeen, Randall Plant and Mary Anne Ross. 1980. Fuel Alcohol from Grain: Energy and Dollar Balances of Small Ethanol Distilleries and their Economies of Size and Scale. Staff Paper 80 E-151. Department of Agricultural Economics, University of Illinois, Urbana. (Also issued as ERG Document No. 313, Energy Research Group, Office of Vice Chancellor for Research, University of Illinois.)

Dovring, Folke and John F. Yanagida. 1979. Monoculture and Productivity. AE-4477. Department of Agricultural Economics, University of Illinois at Urbana-Champaign.

Hannon, Bruce F. and Horacio Perez Blanco. 1979. Ethanol and Methanol as Industrial Feedstocks. ERG Document No. 268. Energy Research Group, Office of Vice Chancellor for Research. University of Illinois at Urbana-Chamgaign.

Herman, Marilyn. 1981. Food and fuel: raw material availability for the United States' alcohol fuels program. Presented at the AAAS Annual Meeting, Toronto, January 5.

Illinois Environmental Protection Agency. 1979. Water Quality Management Plan. Volume 3: Nonpoint Sources of Pollution. State of Illinois, Springfield.

Jantzen, Dan and Tom McKinnon. 1980. Preliminary Energy Balance and Economics of a Farm-Scale Ethanol Plant. SERI/RR-624-669R. Solar Energy Research Institute, Golden, Colorado.

Ladisch, M.R., Christine M. Ladisch and George Tsao. 1978. Cellulose to sugars: new path gives quantitative yield. Science 201:743-745.

Lieth, H. 1972. Modelling the primary productivity of the world. Nature and Resources 8(2):5-10.

Meekhof, Ronald, Mohinder Gill and Wallace Tyner. 1980. Gasohol: Prospects and Implications. Agricultural Economic Report No. 458. National Economics Division, Economics, Statistics, and Cooperatives Service, U.S. Department of Agriculture, Washington, D.C.

Miller, D.R. 1980. Cellulose-based fermentation alcohol. In Proceedings of the First InterAmerican Conference on Renewable Sources of Energy (New Orleans, Louisiana, November 25-29, 1979), pp. 175-182. Published by First InterAmerican Conference on Renewable Sources of Energy, New Orleans.

Office of Technology Assessment. 1980. Energy from Biological Processes. U.S. Congress, Washington, D.C.

Schnittker Associates. 1980. Ethanol: Farm and Fuel Issues. U.S. National Alcohol Fuels Commission, Washington, D.C.

Solar Energy Research Institute. 1980. Fuel from Farms. SERI/SP-451-519. Golden, Colorado.

Spano, Leo, Thomas Tassinari, Dewey D.Y. Ryu, Alfred Allen and Mary Mandels. 1980. Enzymatic hydrolysis of cellulose to fermentable sugar for production of ethanol. In Proceedings of the First InterAmerican Conference on Renewable Sources of Energy (New Orleans, Louisiana, November 25-29, 1979), pp. 157-174. Published by First InterAmerican Conference on Renewable Sources of Energy, New Orleans.

Tyner, Wallace. 1981. The potential of using biomass for energy. Presented at the AAAS Annual Meeting, Toronto, January 5.

United States Department of Agriculture. 1980. Small Scale Fuel Alcohol Production. U.S. Government Printing Office, Washington, D.C.

Wisner, R.N. and J.O. Gidel. 1977. Economic Aspects of Using Grain Alcohol as a Motor Fuel, with emphasis on By-product Feed Markets. Econ. Rep. Ser. No. 9, Dept. of Economics, Iowa State University, Ames.

William Lockeretz

5. Environmental Implications of Growing Corn for Alcohol Production

The Future Alcohol Fuel Situation and the Role of Corn

Ethyl alcohol, or ethanol, is currently the United States' only non-petroleum liquid fuel produced in significant quantities. The sharply increased interest in alcohol fuel in the mid-1970s, which was especially prominent in the Midwestern grain belt, has begun to show tangible results in the form of a commercial industry. By the beginning of 1981, U.S. alcohol fuel production capacity was about 100 million gallons per year, a five-fold increase compared to mid-1979, and alcohol/gasoline blends ("gasohol") were available at over 9,000 service stations in 45 states (OTA, 1979; U.S. National Alcohol Fuels Commission, 1981). While current capacity still corresponds only to about 0.1% of gasoline consumption, planned additions are expected to increase the total fourteen-fold by the beginning of 1985 (U.S. National Alcohol Fuels Commission, 1981). This development reflects interest on the part of private companies as well as public inducements, especially gasohol's exemption from the federal gasoline excise tax and additional tax incentives of various kinds in 33 states as of 1980.

This growing interest has led to several recent analyses of the broad prospects for alcohol fuels in the near- to mid-term, both by official bodies (OTA, 1979, 1980; DOE, 1979;

Much of the material in this chapter was adapted from the report "Biomass Energy Systems Program Environmental Assessment," prepared under a contract with the U.S. Department of Energy, Biomass Energy Systems Division. However, the views and opinions expressed are those of the author, and do not necessarily reflect the views of the Department of Energy.

Energy Research Advisory Board, 1980; U.S. National Alcohol Fuels Commission, 1981) as well as by individual investigators (e.g., Commoner, 1979; Ferchak and Pye, 1981).

These analyses often diverge sharply. For example, the Energy Research Advisory Board (1980) projects that ethanol production from grain will not exceed about 300 million gallons per year by 1985, whereas DOE (1979) and Commoner (1979) regard 3.8 and 6.0 billion gallons per year respectively as the maximum obtainable from grains in 1985, with the figures rising to 14.5 and 10.0 billion gallons per year if crop residues and sugar crops are included. Most longer term analyses consider that an optimistic but reasonable goal for the year 2000 would be to displace about 10% of current gasoline consumption, which would require about 10 billion gallons of ethanol per year. However, Ferchak and Pye (1981) believe that just with starch and sugar crops production could be as high as 30 billion gallons by then, while Commoner (1979) puts the figure at 50 billion. Both studies claim that if crop residues, other wastes, and wood are included, the entire automotive fuel demand (around 100 billion gallons per year) could be met with ethanol.

There are also differing views regarding the likely scale of ethanol production facilities and the optimal end use. Hohmann (1980) envisions substantial adoption of on-farm production in the Midwest, whereas the Office of Technology Assessment (OTA, 1979) sees small-scale production as beset by serious problems that may limit its adoption. (Dovring, in Chapter 4 of this volume, discusses some of the economic and technical differences between small- and large-scale production.) Regarding end use, most analyses assume that ethanol will be used primarily in a 10%/90% blend with unleaded gasoline, which is the present practice. However, the U.S. National Alcohol Fuels Commission (1981) emphasizes the advantages of straight alcohol, and the Alcohol Fuels Policy Review (DOE, 1979) describes several possible non-transportation uses, such as in boilers, fuel cells, and petrochemical synthesis.

The likely mix of raw materials for ethanol manufacture is a particularly complex question on which recent analyses have diverged sharply. Possible non-fossil sources include forestry (either logging residues or wood raised for the purpose), municipal solid waste, and agriculture. Whatever the comparative merits of these sources, no one by itself could displace all fossil fuel use, at least not in the near- to mid-term. Thus they are not really in competition with each other. For example, whether municipal solid waste will be made into ethanol will be determined by the costs and benefits of doing this, regardless of whether or not wood is also made into ethanol. (However, massive production of methanol from coal, if it can be done cheaply enough, might preempt ethanol production from biomass or municipal waste, since it would not be limited by raw material availability.)

However, when we restrict our attention to agriculture, then the comparative merits of various raw materials do become very relevant, since they compete for the same limited production resources, such as cropland and irrigation water. Ethanol can be made from virtually any agricultural material high in starch, sugar, or cellulose. Corn grain dominates the current situation, but various alternatives have their respective proponents. Commoner (1979) sees a major role for sugar beets, while DOE (1979) projects that sweet sorghum, presently a minor crop, will be the leading raw material by 1990. Forage grasses and other silage-type crops have been described very favorably by OTA (1980) and Dovring (Chapter 4 of this volume). Most analyses also assume that crop residues will eventually become extremely important, although the technology for their conversion is not yet commercially competitive and is receiving intense research attention in the hope of making it feasible.

Regardless of the longer range situation, there is little question that at least for the next several years, and very likely even longer, none of these materials will challenge the supremacy of corn grain. Of materials convertible to ethanol with commercially available technology, only corn is already being produced on a large enough scale to displace a significant amount of gasoline (Table 1). Moreover, along with grain sorghum, to which it is agronomically very similar, it currently is the cheapest raw material when one takes account of by-product credits and conversion costs (Table 2). Not surprisingly, over four-fifths of the ethanol production capacity expected to be in place in 1985 will use corn (U.S. National Alcohol Fuels Commission, 1981). Consequently, this chapter concentrates on corn. Crop residues, the second leading contender as the major raw material in the mid-term, are discussed in the following chapter.

The Overall Environmental Significance of Alcohol Fuels

There are several motivations for replacing petroleum-derived fuels by renewable sources. Since the mid-1970s the most obvious and most publicized have been the sharp price increases, our vulnerability to cutoffs of overseas supplies, and the strain on our balance of payments. But already by

Table 1. Ethanol equivalent of national production of various agricultural materials

Material	Area[a] (million acres)	Production[a] (million tons)	Ethanol Equivalent (billion gallons)
Corn grain	71	198	18
Wheat	62	60	5
Sorghum grain	14	22	2
Sugar cane	0.7	26	0.4
Sugar beets	1.2	24	0.5
Major crop residues	230	440[b]	23

[a] 1977-79 average. From USDA (1980c).
[b] Gross production, with no allowance for harvesting losses and with no consideration of soil conservation requirements, such as those discussed in Chapter 6.

Table 2. Comparative costs of ethanol from various agricultural raw materials

Material	Cost[a] ($/gallon)
Corn grain	$1.05
Sorghum grain	1.02
Wheat	1.31
Sugar cane	2.07
Sweet sorghum	1.40
Corn residue (stover)	1.32
Wheat residue (straw)	2.13

[a] All costs are for 1978, and include raw materials cost, by-product credits (where applicable), fixed and variable operating costs, and 15% return on investment. From DOE (1979).

the early 1970s, when oil imports were below their recent level and the first OPEC price increases had not yet occurred, a strong inducement to find renewable alternatives to petroleum was being provided by environmental concerns. Events such as the oil spill from the Amoco Cadiz have since reaffirmed the validity of this view, even if some earlier advocates of renewable energy may have been slightly naive in failing to anticipate that some high-technology renewable energy concepts would be proposed that could be just as environmentally destructive as fossil fuels. The economic situation of the past few years thus has had a mixed effect on the prospects for environmentally benign energy technologies. On one hand, it has provided a powerful inducement for the adoption of renewable energy systems, but it also has meant that a renewable system need not be environmentally attractive to be adopted, and that those that are might find themselves pushed aside by those that aren't.

Thus the environmental significance of alcohol fuels, like that of any other renewable energy concept, must be regarded as an open question. This chapter deals with only one component of that question, namely production of corn as the raw material. The overall environmental impact of alcohol fuel also includes conversion of the raw material, transportation, and end use. However, these three areas are considerably less complex, and are not likely to involve significant adverse effects. Alcohol spills are expected to be less environmentally damaging than those of petroleum, with their effects not nearly as long-lasting (OTA, 1980). Alcohol production facilities will generate airborne and waterborne wastes similar to those of many other industrial facilities and controllable with available technology (Argonne National Laboratory, 1980). (Small-scale facilities, whose boiler emissions are not readily controlled, are an exception, however.) End use air pollution effects are expected to be less severe than those of gasoline in some ways, but more severe in others, assuming the alcohol is used in a blend with gasoline. However, the adverse effects (higher evaporative hydrocarbon and aldehyde emissions) can be mitigated if new car designs specifically take gasohol into account and if gasoline is refined to optimal specifications for blending with alcohol (OTA, 1980). While further study of the quantitative details is needed (mainly in connection with end use effects), in general the environmental problems from production, transportation, and use of alcohol fuels seem manageable and qualitatively similar to those already encountered with fossil fuels.

This is not true for the production of the raw material, however, even if it is corn, a familiar crop (in fact, the nation's leading crop). Because agriculture is so much more diffuse than most other production processes and because it is sensitive to so many external variables, such as weather and soil type, its environmental effects are very difficult to regulate or even to measure. Moreover, many of its most significant effects are indirect or not apparent except in the long term. As noted in the introductory chapter, even a moderately ambitious alcohol fuel program based on corn would cause major shifts in the current agricultural system. Such a program, in consequence, could have some very significant environmental effects.

This chapter examines some of those effects. Because of the highly variable nature of agriculture, no attempt is made to present comprehensive quantitative conclusions. Rather, only the basic concepts are discussed, illustrated as appropriate by representative numerical examples that are intended to be no more than that.

What Do We Mean by the Environmental Effects of Corn Production for Alcohol?

In assessing the environmental significance of a corn-based alcohol program, it first is necessary to specify what environmental effects are involved, a question that is considerably more complex than it might seem at first. It is sometimes assumed that corn production would increase by the amount consumed by alcohol production, since the desired level of alcohol production is often taken as the maximum that would not take corn from any other users. Second, it is sometimes assumed that since corn used for alcohol is indistinguishable from corn raised for other purposes, the environmental impacts of producing a unit of it for alcohol can be taken as the "average" environmental impact per unit of corn production in general. (An exception to this simplified view is that "new" corn land is usually distinguished from "old" corn land, since in this case the incremental and the previous production are physically separate.) Both of these assumptions are highly oversimplified and can seriously distort the environmental significance of alcohol production.

The total quantity of corn produced reflects the combined decisions of several hundred thousand independent producers reacting to the price of corn and of alternative crops, the prices of production inputs, and non-economic factors such as personal preferences. It is not determined by a central agency, except to the extent that land may be intentionally idled under federal commodity programs. (From

1974 to 1981, corn land was intentionally idled only in 1978 and 1979). Because corn is an interchangeable commodity sold into a single market, its price is the same to all users, whether they buy it for export, feed, food, or alcohol production. Except under certain circumstances, such as an export embargo, or when prices fall to near the government support level, market forces determine the price of corn and thus its distribution among these different end uses.

This means that turning a given amount of corn into alcohol does not automatically increase corn production by that amount. Rather, the alcohol program adds to the demand for corn and thereby increases its price. However, price increases will lead not only to increased production, but also to decreased consumption for other uses. Consequently, only some of the raw material for alcohol will come from additional production; the remainder comes from foregone consumption for other purposes. The exact distribution between these sources depends on the price elasticities of supply and demand and can only be predicted, if at all, through fairly complicated economic models. One such model was used by Meekhof et al. (1980) to estimate the consequences of using 769 million bushels (20 million metric tons) of corn to produce 2 billion gallons of ethanol annually by 1985. They predicted that the price of corn would increase $0.22/bu ($9/metric ton), or about 8%, and that production would increase by 331 million bushels. Non-alcohol domestic use was predicted to drop by 101 million bushels and exports by 153 million, with the remainder of the needed corn coming from a drawdown of stocks.

While the precision with which Meekhof et al. stated their results undoubtedly overstates the reliability of this type of prediction, nevertheless some increase in corn production will result from an alcohol program. In addition, there will be other changes in crop mix and production practices. Unfortunately, one cannot identify the environmental impacts of that portion of corn production specifically destined for alcohol production, in contrast, for example, to the straight-forward identification of the air emissions from a plant that produces alcohol. Rather, the environmental impact is the difference (however one "subtracts" one level of environmental impact from another) between the total environmental impact of the agricultural system with and without alcohol production. Thus two steps are involved. The first is to determine how the agricultural system responds to a given increase in the demand for corn (in this case for alcohol, although similar computations could be used for any new use of corn). The second is to evaluate the environmental impact of this change. Only the

Table 3. Indexes of crop output, inputs, and output per unit of input, 1940-1979[a]

	Crop Output	Crop Production Inputs					Output per Unit of Input	
		Cropland	Labor	Chemicals	Machinery	Irrigation[b]	Land	Labor
1940-44	71	109	305	18	50	52	65	24
1945-49	77	111	257	23	66	66	70	30
1950-54	79	112	206	34	92	75	71	38
1955-59	84	107	159	43	97	85	79	54
1960-64	95	110	129	59	94	95	94	72
1965-69	100	99	101	95	98	100	102	100
1970-74	111	101	87	129	103	105	110	128
1975-79	129	109	82	154	121	131	118	159

[a] From USDA (1981). All indexes relative to 1967 = 100.

[b] Irrigated area in last year of indicated period. From USDA (1979).

latter will be discussed here. Examples of the first part of the analysis include the previously cited one by Meekhof et al. (1980), as well as those of Wisner and Gidel (1977), Hertzmark et al. (1980), and Schnittker Associates (1980).

The distinction between environmental impact of a given amount of "average" corn production and that of the same increment to total production is an important one because many agricultural processes are highly non-linear. In particular, the environmental impact of an increment to total production is likely to be considerably greater than the average. For example, one way of increasing total production would be through higher fertilizer applications. As will be discussed below, the fraction of the fertilizer application that potentially could pollute ground or surface waters increases rapidly as the fertilization rate increases. The quantitative details of this type of relationship vary sharply from one location and cropping system to another. Thus the general discussion that follows must be regarded as merely an introduction to the entire topic, and any attempt to develop national estimates of the environmental impact of raising corn for alcohol production must involve a large amount of site-specific data.

Possible Sources of Increased Corn Production

Some background to the question of how corn production might be increased to meet the alcohol fuel demand is provided by the historical record, specifically the period since World War II, during which crop output increased sharply. Table 3 shows that from the early 1940s to the late 1970s crop production almost doubled, although the amount of cropland remained constant and the use of labor fell almost fourfold. On the other hand, there were sharp increases in other major inputs, such as machinery and irrigation, which more than doubled, and chemicals (fertilizers and pesticides), which increased almost ninefold. Irrigation primarily increases yield, and machinery primaily reduces labor, while chemicals do both. For example, if fertilizers are used on land where no nutrients were applied previously, then output increases, but when they replace an equivalent quantity of manure (a much bulkier material that is harder to handle and apply), then labor will decrease while yield might not change. Likewise herbicides could increase yield if they can control a previously intractable weed problem, but they might instead provide the same degree of control, with the same resulting yield, if they replace effective but labor-intensive mechanical cultivation or hand weeding.

For corn production in particular, rather than crops in general, data on recent trends are shown in Table 4. The average annual increase in output of about 4% per year has come from two roughly equal sources: increases in yield per unit area, and increases in the total amount of corn land. The latter in turn came mainly from an increase in total cropland, although there also was a shift from other crops to corn. (In Table 4, the data on cropland refer to the main corn-producing regions, which is why the increase is greater than for the national data shown in Table 3.) Table 4 also shows one projection of changes over the next decade, made in connection with a national review of the current alcohol fuel situation and its future prospects. Total corn production is expected to rise more slowly, with increases in corn land and in yield per unit area continuing to be equally important. However, most of the new corn land is projected to come from a shift from other crops.

Table 4. Sources of increases in U.S. corn grain production, past and projected.[a] (Average annual rate of change, in %)

		Increase in Harvested Area				
From	To	Increase in Total Cropland	Shift from Other Crops	Total	Increase in Yield per Unit Area	Increase in Total Production
Past						
1965-69	1970-74	0.9	0.7	1.6	1.4	2.9
1970-74	1975-79	2.1	0.6	2.7	2.5	5.3
1965-69	1975-79	1.5	0.7	2.2	1.9	4.1
Projected						
1978-79	1985	0.1	1.3	1.4	0.7	2.1
1985	1990	0.1	0.3	0.5	1.8	2.2
1978-79	1990	0.1	0.9	1.0	1.2	2.2

[a] Historical data from Lockeretz (1980). 1985 and 1990 derived from base case projections in Schnittker Associates (1980); no effort is made to increase corn supplies for the sake of alcohol production. Total may not equal sum of separate contributions because of rounding.

In relating historical data and projected trends to alcohol production in particular, it is important to distinguish two changes: those that would occur anyway over time (as shown in Table 4, which does not reflect any specific efforts to increase corn supplies for the sake of alcohol production), and those that occur at a given time because of the higher crop prices resulting from an alcohol program. Some projections of the latter type are shown in Table 5. (Note that the increase in corn production is less than that needed to produce the indicated quantity of alcohol, since other consumption will decline, as was discussed previously.) The shift of land from soybeans to corn is particularly important with alcohol production because of the large amount of high-protein by-product (distillers dried grains) resulting from alcohol production, which will substitute for soybean meal (see Chapter 4). However, there will also be some additions to total cropland. The particular study summarized in Table 5 projected only a small increase in yield per unit area (that is, in a given year as a result of an alcohol program), which presumably would result from farmers' slight intensification of practices on existing corn land in response to higher corn prices. However, projections of this sort are highly speculative, as shown by the wide divergences among the studies cited earlier in this chapter. Because corn prices are so volatile and unpredictable even without an alcohol program, it is possible that an alcohol program could increase prices by considerably more than was projected, and that farmers will therefore be motivated to intensify their practices substantially. Because of the non-linearities discussed earlier, this could have appreciable environmental consequences. Therefore this chapter will discuss all three of the possible sources of increased production: shifts from soybeans to corn, shifts to corn from non-cropland uses, and intensification of existing corn production. Whether the relative importance of these three sources will actually agree with the projections of Table 5 is, of course, completely conjectural.

Shifting Land from Soybeans to Corn

The shift of land from soybeans to corn because of a great increase in the supply of high-protein feeds is generally regarded as the most important source of new corn production with an alcohol fuel program. Such a change is made feasible by the fact that the nation's leading corn and soybean producing regions overlap considerably. Of the eight leading soybean states, the six that are entirely or partially in the Corn Belt (Illinois, Iowa, Missouri, Minnesota, Indiana, and Ohio), which together account for over half of the total soybean area, also have almost three-

Table 5. Projected changes in corn production for various levels of ethanol production[a]

	Ethanol Production (bil. gal.)	Corn Land (mil. A)		Yield (bu/A)	Production (bil. bu.)	Source of New Corn Land[c] (mil. A)	
		Planted	Harvested[b]			Former Soybean Land	New Cropland
1985							
	1	88.0	77.4	110	8.52	--	--
	2	89.5	78.8	110	8.67	1.5	0
	4	95.0	83.8	111	9.30	6.0	1.0
1990							
	1	90.0	79.2	120	9.51	--	--
	2	92.5	81.4	120	9.77	2.0	0.5
	4	97.0	85.6	121	10.36	6.0	1.0

(a) Based on Schnittker Associates (1980). All ethanol assumed to be produced from corn (2.5 gal/bu).

(b) Harvested for grain. Remainder primarily is land harvested for silage.

(c) Relative to the same year's "base case" of 1 billion gallons of ethanol per year. This is the quantity projected as available from corn without increasing the real price of corn, with no set-aside program, and with no special measures to increase corn production.

fifths of the total corn area even though they have only three-tenths of the nation's cropland (Lockeretz, 1980). Many farms in this region already specialize in these two crops, with the mix shifting back and forth from year to year in accordance with their relative prices. Thus a shift from soybeans to corn would not represent a radical change in cropping practices. The magnitude of such a shift would be limited, however, by potential insect problems and by the fact that too great a specialization in corn could create an excessive peak labor demand and would mean that expensive machinery is used for only a very short time each year. (Periods of peak labor demand and machinery use for planting and harvesting differ slightly between corn and soybeans.)

Such a shift would have three main environmental consequences: a reduction in soil erosion, an increase in the use of nitrogen fertilizer, and an increase in the use of insecticides. In the Corn Belt the two crops are commonly grown in alternate years, i.e., the well-known corn/soybean rotation, or in longer rotations, such as corn/corn/soybeans or various rotations with hay and oats. Thus one important way of shifting land from soybeans to corn will be to replace corn/soybeans with continuous corn. A direct comparison of these two systems showed that for a particular location in Iowa, over the two-year rotation annual soil loss was about one-eighth lower with continuous corn than with corn/soybeans (Laflen and Moldenhauer, 1979). In other rotations as well, erosion with soybeans typically is greater than with corn, since soybeans produce less residue and provide less complete ground cover during the growing season (Wischmeier and Smith, 1978). Of course, the erosion-reducing effect of switching from soybeans to corn would be lost if the corn residues are removed for alcohol production or other energy uses, as will be described in Chapter 6.

Because soybeans are a nitrogen-fixing legume, they typically receive little or no nitrogen fertilizer. In contrast, corn is the nation's leading consumer of nitrogen fertilizer. In Illinois, for example, the average application rate is about 150 lb/acre. (The yield response of corn to nitrogen was discussed in some detail in Chapter 2). Applications of phosphorus and potassium are also higher on corn than on soybeans, but not by nearly as much as in the case of nitrogen. Nitrogen fertilizer is of special significance because it is highly mobile and can contribute to elevated nitrate levels in drinking water, a potential health hazard. Switching from corn/soybeans to continuous corn, besides doubling total nitrogen applications, will also eliminate the "scavenging" effect of soybeans on the nitrogen that remains in the soil following corn, which helps reduce the leaching of nitrogen to groundwater (Welch, 1979). The loss of nitrogen to the environment is discussed in more detail later in this chapter.

Finally, switching from soybeans to corn will greatly increase the use of insecticides. Insecticides were applied to 39% of corn land in the Corn Belt in 1976, but to only 1% of soybean land (Eichers et al., 1978), and the total quantity of active ingredients applied on corn was 120 times greater, even though there was only 1.6 times as much land in corn as in soybeans. Actually, switching from corn/soybeans to continuous corn will result in an increase in insecticide use that is even greater than what would be calculated purely on the basis of current application rates per acre. A major reason for using insecticides on corn is to control corn rootworm. This is a serious problem only with continuous corn, so that insecticides sometimes are not used on corn in rotation. Therefore, while insecticides will be used virtually every year with continuous corn, they would have been used only in some corn years and virtually no soybean years with corn/soybeans. Thus insecticide use more than doubles when land is shifted from corn/soybeans to continuous corn. This illustrates the non-linearities of environmental impacts discussed earlier and shows the difference between average or existing production compared to new, incremental production.

New Cropland

How much the total amount of cropland will increase with an alcohol program, and what the effects of such an increase will be, is a particularly complicated and disputed question. The analysis of Schnittker Associates (1980) discussed earlier (Table 5) predicted that cropland would increase by only 1 million acres (0.4 million hectares) as alcohol production from corn increased from 1 billion to 4 billion gallons per year (4 billion to 15 billion liters per year). This analysis projected most of the additional corn to come from former soybean land. In contrast, another analysis (Tyner, 1980) predicted an increase of 2 million acres for each additional 1 billion gallons per year, which is six times the rate given by Schnittker et al. One uncertainty affecting calculations of this kind is the extent to which high-protein by-product feeds from alcohol production can substitute for soybean meal when they are available in large amounts (in contrast to the current situation, in which the relatively small by-product feed supply from fuel and beverage alcohol production is readily absorbed by cattle producers).

A further complication is that the land which might be brought into crop production is very non-homogeneous with regard to both its expected productivity and the ease with which it can be put into crops. For example, a 1975 U.S. Department of Agriculture survey (Lee, 1978) identified 111 million acres (45 million hectares) as having high or medium potential for conversion to cropland. But only 78 million acres of this was regarded as having high potential, that is, a high probability of being converted to cropland within 10 to 15 years. Of this, only 35 million acres was regarded as not having problems that would inhibit its development as cropland, and only 24 million acres was considered to be in the prime productivity category. In other words, any analysis based on what is claimed to be the amount of potentially croppable land necessarily is highly oversimplified.

More than just the physical suitability determines what will actually happen to a piece of land under a particular set of circumstances. Some potential cropland requires long-term capital investment, such as drainage. For a landholder to be willing to undertake the necessary expenses, not only must current economic conditions (land values, crop prices, and production costs) be perceived as favorable, but also the long-term prospects. These factors are highly volatile, as shown by the fact that from 1975 to 1977, the amount of land regarded as having "high" potential for conversion fell by one-half from the earlier value of 78 million acres (USDA, 1980a), mainly because crop prices had fallen from the record values reached in 1974. Moreover, there are a great number of factors besides economic profitability that may hinder development of potential cropland for cropping. For example, it might be held for speculation and eventual conversion to non-agricultural uses, or the owner may have inadequate capital to make the needed improvements. Lee (1981) reported that 43% of the high and medium potential cropland in the 1977 survey was affected by obstacles of this kind.

The heterogeneity of potential cropland has important environmental implications. Table 6 shows the breakdown by soil capability class and subclass. Suitability for cropping decreases with increasing class number, with Class I having virtually no limitations, while Classes V through VII are considered suitable only for grazing or other non-crop uses and Class VIII not even for grazing. Within each class, the subclass e has erosion as its primary limitation, whereas land in other subclasses is limited mainly by problems such as poor drainage, rocky or thin soil, or unsuitable climate. The "potential cropland" category shown in Table 6 primarily consists of rangeland, unimproved native hay and pasture, and

Table 6. Distribution, by land capability class, of high and medium potential cropland, present cropland not in intensive use, and cropland in intensive use, 1977[a]

Capability Class[b]	Land Not in Intensive Use				Cropland in Intensive Use	Average Erosion in Intensive Use (tons/acre-yr)
	Potential Cropland		Present Cropland	Total		
	High	Medium				
Few limitations						
(Class I)	5.5%	1.5%	4.2%	3.1%	9.0%	3.2
Erosion-limited						
(Subclass e)						
II	26.3	14.8	20.8	18.9	22.8	5.6
III	19.7	26.7	22.3	23.8	18.1	9.6
IV	4.7	12.3	10.1	10.0	6.1	15.0
Other limitations						
(Subclasses other than e)						
II	20.7	11.3	17.5	15.1	26.7	3.2
III	12.0	13.4	11.6	12.6	12.6	3.2
IV	4.0	7.6	5.6	6.3	2.8	3.6

Table 6, continued

Capability Class[b]	Land Not in Intensive Use				Cropland in Intensive Use	Average Erosion in Intensive Use (tons/acre-yr)
	Potential Cropland		Present Cropland	Total		
	High	Medium				
Not suited for cropping						
(V through VIII)	7.2	13.0	7.9	10.2	2.0	--
Total area (million acres)	40.1	94.9	56.4	203.2	104.9	--

(a) Intensive use is row crops or close-grown cultivated crops. Cropland not in intensive use generally is in rotation hay or pasture. Potential cropland primarily includes native pasture, rangeland, and forests. Derived from data in OTA (1980).

(b) Limitations on cropping become more severe with increasing capability class. Non-erosion limitations include wetness, soil problems, and climate.

non-agricultural land such as forests. The table also shows another potential source of increased corn land, namely cropland managed in other ways than continuous cropping, such as hay or pasture in rotation with cultivated crops. (This is cropland that didn't happen to be in cultivated crops for the particular year that the data refer to.) Finally, for comparison, the table also shows the capability class breakdown for land that currently is in intensive use, i.e., row crops such as corn and soybeans and close-grown crops such as wheat. The last column shows the average erosion rate by capability class for this land use.

Table 6 shows that the land that could be cropped more intensively is more likely to have an erosion problem than land already in crops. For example, of land that has either a high or medium potential for conversion to cropland or that already is cropland but not in intensive use, one-third (33.8%) is in capability subclasses for which erosion is a severe or very severe limitation (IIIe and IVe), whereas for land being cropped intensively, the corresponding fraction is only one-fourth (24.2%). Land in subclasses IIIe and IVe under intensive cropping is eroding at two to three times the so-called soil loss tolerance, which typically is set at 5 tons/acre-year (11 metric tons/hectare-year). This is the level considered compatible with sustained productivity, although it is considered to be too high by some researchers.

All that Table 6 shows is what land could be put into cultivation, not what will happen. The choice is still up to the individual landholders. If they restricted themselves to the "high potential" category, then the erodibility of new and old cropland would not be very different (compare columns 1 and 5). On the other hand, since such decisions are not regulated, and since so many factors enter--erosion being just one, and not necessarily the most important one--it is possible that economic pressures could encourage cropping of seriously erodible land. If so, then this would be another example of how an adverse environmental impact may increase faster than proportionally to the amount of corn produced.

Unfortunately, the historical record suggests that the latter might occur. Several times during the past century sharp crop price increases from wars or other causes have led to sudden increases in the area in cultivated crops. Each such increase was followed in turn by periods of severe erosion, most spectacularly the great Dust Bowl of the 1930s, whose complicated causes included the rapid increase in grain prices during and immediately following World War I (Lockeretz, 1978). The most recent example was the 1974 call for farmers to plant "fencerow to fencerow" to achieve "all-out food production." It is irrelevant whether their positive response to this call was prompted by a concern for the "world food crisis" or by the record high crop prices at the time. In any case, a total of almost 9 million acres was converted to crops in 1974, almost half of which had inadequate erosion control. On all the new cropland, erosion averaged about 7 tons/acre-year, while on the 4 million acres with inadequate erosion control, the average was 12 tons/acre-year (Grant, 1975).

Besides representing a loss of an important natural resource, soil erosion also causes water quality problems through sedimentation, pesticide contamination, and increased nutrient loads. Many of the most effective methods for controlling these forms of water pollution are the same as the major soil erosion control methods (Stewart et al., 1976). Conversely, increasing erosion through an expansion of intensive cropping to inappropriate lands will increase all of these water pollution problems.

Intensified Production on Existing Corn Land

Nitrogen Fertilization

Using more nitrogen fertilizer is one of the easiest ways for a farmer to intensify corn production in response to higher corn prices. It also is an environmentally important change. Even though the economically optimal nitrogen fertilization rate will change only slightly for any reasonable change in either the price of corn or the price of nitrogen (as was shown in Chapter 2), nevertheless, there will be a major change in the amount that is not taken up by the crop and that is a potential water pollutant. (Whether the amount actually ending up in the water increases proportionally depends on a great many site-specific factors, since nitrogen not taken up by the crop can follow many different pathways, not all of which result in water pollution.)

This is illustrated by the numerical data on corn yield response to nitrogen already presented in Table 3 of Chapter 2. Suppose, for example, that initially the price of corn and of nitrogen were such that the economically optimal fertilization rate was 160 lb/acre. Since one bushel (56 lb) of corn at 1.4% nitrogen content (a typical value) contains 0.8 lb of N, this means that 68 lb of N are contained in the 85 additional bushels (133 - 48) produced by applying 160 lb of N. Consequently, 92 lb of N may be lost to the environment, or 1.1 lb for each additional bushel of production. (Actually, since the N content of corn is lower at low N application rates, the N content of the crop increases by more than 68 lb. In addition, some of the applied nitrogen goes to compensate for the loss of soil nitrogen taken up by the first 48 bushels produced in the absence of fertilizer. Both of these effects, ignored here in the interest of simplicity, would strengthen the following argument, since they mean that in fact less than 92 lb of N is not taken up by the crop under the initial application rate of 160 lb/acre.)

Now suppose that in response to an increase in the price of corn (and no corresponding increase in the price of N, a highly optimistic assumption in view of the behavior of fertilizer prices during the sharp crop price increases of the early 1970s), the fertilization rate is increased to 180 lb/acre, producing 1 additional bushel. Of the last 20 lb of N, only 0.8 lb is taken up by added production, and 19.2 lb per additional bushel is available for loss to the environment. In contrast, for the initial 160-lb application, the figure was only 1.1 lb. Putting it another way, increasing the yield by only 1 bu/acre (from 133 to 134) results in a 20% increase in the unused nitrogen fertilizer (from 92 to 111 lb). This is a particularly dramatic example of a non-linearity in environmental effects (or, more correctly, in potential environmental effects, since we cannot say for sure where the unused nitrogen will end up.) Thus it further illustrates the importance of specifically examining incremental changes in agricultural production caused by an alcohol fuel program, rather than simply extrapolating on the basis of averages for the current situation.

Irrigation

The expansion of irrigation in the Great Plains has led to a dramatic change in the region's agriculture, especially in the more arid lands in the western portion of the region, where it has permitted intensive cropping of land formerly used mainly for grazing. (In the eastern portion, irrigation often is a supplementary source of water, and non-irrigated cropping also occurs, although with lower yields and more risk of crop failure because of drought.) In Nebraska, the nation's third leading corn-producing state in most years, irrigation accounts for over three-fourths of total production (Nebraska Department of Agriculture, 1979). Nebraska, Kansas, and the adjacent High Plains portion of Colorado accounted for 93% of the total increase in irrigated land between 1964 and 1974, with the High Plains portions of these three states alone accounting for 42% (Frederick, 1980).

Most of this increase has been achieved by using groundwater, especially with center pivot systems, which consist of a one-quarter mile boom that slowly sweeps a 132-acre circle. Center pivot systems are suitable on rolling terrain, since the boom is articulated and therefore can conform to the topography. As with other sprinkler irrigation systems, they can be used on very sandy soils on which too much water would be lost with furrow irrigation. Consequently center pivots have been responsible for the expansion of crop production into the Sand Hills of Nebraska, an area once considered suitable only for low-density grazing. The source of groundwater for High Plains irrigation is the Ogallala aquifer, an enormous underground reservoir that unfortunately recharges at a rate far below current withdrawals for irrigation. Depletion of the water table has already been a problem, especially in the Southern Plains (Frederick, 1980), and is expected to become serious in the near- to mid-term in some corn regions further north. The speed with which this will occur is highly variable from place to place, however, and some areas will continue to have water for a few decades.

But even where the water situation is adequate for the near future, it is very doubtful whether the boom in center pivot irrigation in the Plains, which began in the 1960s and accelerated in the early 1970s, will be able to continue (Lagrone, 1979). Because of the high pumping depths and the high pressure required for water distribution (higher than most other irrigation methods), center pivot irrigation is extremely energy-intensive. For example, corn production in Southwestern Nebraska requires over 900 kwh of electricity per acre (FEDS, 1981), or equivalent amounts of natural gas or diesel, which also are common pumping fuels. This amounted to about 13% of the value of the crop in 1979. Problems in meeting peak electricity demand have also limited the growth of this technology (Sanghi and Johnson, 1978).

But regardless of what other changes would have occurred, an increase in the price of corn, for example, as a result of an alcohol fuel program, will increase corn production under center pivots in the Plains and speed the drop in the water table compared to what the situation would have been otherwise. (This might mean that the rate of decline in irrigation is slowed, rather than that the rate of growth is increased.) This increases future pumping costs and hastens the eventual economic depletion of the groundwater resource (the point at which it no longer is economical to irrigate). Because the forces that determine the amount of irrigation are extremely complex and largely unpredictable, it is impossible to make any quantitative statements about this effect.

Irrigation may also result in N pollution of groundwater, especially on sandy soils with low capacity for retaining N and consequently a high susceptibility to leaching. (Recall that this is the very characteristic that makes sprinkler irrigation the preferred method.) Irrigated corn was identified as the likely major contributor to groundwater nitrate levels in central Nebraska that exceeded the recommended public health limit (Spalding et al., 1978; Gormley and Spalding, 1979).

As far as erosion is concerned, irrigation can have a beneficial effect on the Great Plains in that it insures a crop that is adequate to protect against wind erosion. With dryland farming, in contrast, a drought could result in a crop failure and thereby leave the soil exposed and susceptible to blowing. Expansion of crop production to areas of the Plains with marginal natural moisture was the main reason for the terrible dust storms during the 1930s drought, a phenomenon repeated less spectacularly during the 1950s and 1970s. On the other hand, some center pivot systems have been installed on land that previously was in pasture, which provides excellent erosion control. Some of this land is extremely erosion prone in the absence of a cover, and there have been problems with "blowouts" on land that was inadequately protected after the growing season.

Elimination of Intentionally Idled Land

The discussion thus far has been concerned with responses to an alcohol program under free market conditions, which is what has prevailed since 1974, with the exception of 1978 and 1979. However, during the 1960s and early 1970s, a considerable amount of land was intentionally idled under federal programs intended to support the prices of feed grains (mainly corn and sorghum), wheat, and other major crops. Sometimes as much as about 40 million acres was taken out of feed grain production alone. Not surprisingly, when interest in alcohol fuel from grain revived in the 1970s, one of the most attractive and most frequently heard arguments was that an alcohol program would provide a market for the grain from "surplus" land. Thus grain could be used for alcohol production without having to take it from other uses, and the government could end its price support payments, or use them instead to support alcohol production.

But even though crop surpluses and price supports were so common during most of the post-World War II period that many people came to regard them as inevitable, it is not clear whether we can continue to count on excess production capability. To some people, the sharp world-wide grain price increases and short supplies of the mid-1970s signalled a permanent change to a new era of shortages rather than surpluses. To others, these shortages were a temporary anomaly, a view that got at least short-term corroboration from the falling prices of the late 1970s and a return to the set-aside program for two years. It is difficult to know yet which of these two views will prove correct over the next several years. At the very least, we know that the shortages of the mid-1970s were not as temporary as some people thought. This was shown by the effect, or really the non-effect, of the suspension of grain sales to the Soviet Union in January, 1980. Even though some people feared that this would severely depress prices, in fact total overseas demand remained very strong, and it was decided not to idle any land in 1980. As it happened, this was an extremely fortunate decision: a severe drought during the 1980 growing season sent prices far above the level that the government would have aimed to achieve through a support program.

Regardless of how the long-range picture turns out, it is clear that the amount of "surplus" land in any one year is too variable for such land to be the main source of raw material for an alcohol program. Once the capital has been committed for facilities to convert corn to alcohol, it will not be possible to turn alcohol production on or off according to the availability of agricultural surpluses in a particular year. (This does not apply to systems that are owned and operated by grain farmers, as was discussed in Chapter 4.) Moreover, once the energy system has come to depend on alcohol for part of the liquid fuel supply, turning that supply off would be just as disastrous as a corresponding cutoff in imported petroleum supplies. Nevertheless, in those years in which land otherwise would be idled because surpluses were expected, elimination of intentional idling would obviously be the first way of providing the raw materials for an alcohol program. Consequently, the environmental effects of not idling land must be included among the potential effects of an alcohol fuel program.

At the very least, eliminating the intentional idling of a certain amount of corn land will mean that erosion on it will go from near zero to the average for corn land. This is because under price support programs, idled land must be given a vegetative cover or some other conservation treatment, and management methods are readily available to control erosion on uncultivated land almost perfectly. But the erosion incurred when this land is put into production may be even greater than average. Farmers have considerable freedom as to whether they will participate in the program at all, and if so, which land they choose to idle. Consequently, it may be expected that the less productive land will be idled. Since less productive land generally--but certainly not always--is more erodible, intentional idling can have soil conservation benefits that are more than proportional to the amount of land involved. An analysis of the 1978 set-aside program (Dovring, 1979) concluded that idled land was <u>not</u> less productive than land kept in production. However, county-level data, in contrast to the statewide data used by Dovring, compiled in 1967 showed that in the Corn Belt 35% of land in "conservation use" (which primarily means intentional idling) was in the high erosion categories (soil capability subclasses IIIe through VIIe), whereas for land in crops the corresponding figure was only 22% (Lockeretz, 1980). Because Dovring's analysis is far more recent, however, his conclusion may be more accurate for current programs.

The elimination of the set-aside program in 1974 and the return to "all-out food production" led to very serious erosion problems, as was already discussed. The additional land put into production included about 5 million acres that had previously been idled intentionally, plus 4 million newly put into crops. Thus the high erosion that occurred in 1974, as described earlier, can serve equally well as a warning against both the elimination of intentional idling and inappropriate conversion of land to cropland from other uses.

Policy Implications

This chapter has shown the complexities inherent in trying to anticipate the environmental effects of a corn-based alcohol program. As a result it is not possible to give a simple, overall statement regarding whether such a program would be environmentally acceptable. Too much depends on decisions to be made by farmers, consumers, and policy makers.

For example, a substantial degree of erosion control can be achieved with reduced and minimum tillage methods. But while these methods are being adopted by many farmers, they may create other difficulties, such as aggravated weed or insect problems, and a farmer has to value erosion control sufficiently to be willing to make the change. Greater use of these methods could alleviate the excessive erosion that otherwise might occur with a major alcohol fuel program. However, even without a large amount of alcohol production, excessive erosion is a serious national problem already (USDA, 1980a). What farmers could do and what they will do are not necessarily the same thing. Choice of tillage methods and other soil conservation strategies is a matter that is left to the individual farmer, although the government provides certain inducements and assistance. What the actual erosion situation will be with an alcohol fuel program therefore will also depend on whether the government is willing to strengthen its soil conservation efforts--and perhaps even introduce non-voluntary methods--even though this might work against an all-out alcohol production effort by causing highly erodible land to be kept in soil conserving uses rather than being put into grain production.

In the past few years, there has been a growing interest in non-chemical methods of fertilization and pest control. Even organic farming, a completely non-chemical farming system that only a few years ago was largely regarded as the domain of faddists, recently received a serious examination by the U.S. Department of Agriculture (1980b). For farmers who want to remain closer to their current production practices, there have been some important advances in more efficient fertilizer applications, better manure management, and biological pest control. A well-planned, long-range research program in these areas may lead to further progress that could offset the increase in fertilizer and pesticide use that otherwise would probably occur with an alcohol fuel program, with the resulting environmental impacts discussed earlier.

Consumer preferences are also highly relevant. Most of the corn produced in the United States is used to feed hogs and beef cattle. The early 1970s saw the first reaction against the long established trend of increasing red meat consumption, a reaction prompted in part by sharp price increases and in part by health concerns. It is hard to know what the long-range consumption patterns will be. If consumers show a greater preference for less highly fattened beef, or for more protein from vegetable sources, then alcohol production would not necessarily require an increase in corn production. But here too what could happen and what will happen are two different matters. Fortunately diet is still--in theory--a matter of individual choice, although the economists' ideal of "consumer sovereignty" is far from being fulfilled.

Thus the environmental effects of alcohol production from grain ultimately depend not only on purely technical factors, but also on farmers' and consumers' preference, on agricultural research, and on the government's agricultural and resource policies. This illustrates the basic point made in the introductory chapter, namely, that the energy aspects of agriculture can be analyzed only as part of a much broader analysis of the entire agricultural system. As noted earlier, the scope of agricultural research and of public discussion of agricultural policy has opened up considerably in recent years. This development offers some hope that alcohol production will be carried out in an environmentally sound way. But we can be sure that this will not happen if the issue is approached in too narrow a way, that is, if alcohol is regarded as just another energy source, for which the only important question is "How much can we get, and how soon?" Raising grain for alcohol production is something with a definite market value, where the rewards are direct and immediate; the same cannot be said about preserving environmental values. Without explicit action to establish some kind of balance, the market wins.

References

Argonne National Laboratory. 1980. Environmental Implications of Accelerated Gasohol Production, Preliminary Assessment. ANL/ES-91. Gasohol Task Group, Energy and Environmental Systems Division, Argonne, Illinois.

Commoner, Barry. 1979. Testimony before the U.S. Senate Committee on Agriculture, Nutrition, and Forestry, Subcommittee on Agricultural Research and General Legislation, Hearings on "The Potential for Energy Production by U.S. Agriculture," July 23.

DOE. 1979. The Report of the Alcohol Fuels Policy Review. DOE/PE-0012. U.S. Department of Energy, Assistant Secretary for Policy Evaluation, Washington, D.C.

Dovring, Folke. 1979. Cropland Reserve for Fuel Production. Series E, Agricultural Economics, 79 E-68. Department of Agricultural Economics, University of Illinois, Urbana-Champaign. In The Report of the Alcohol Fuels Policy Review: Raw Material Availability Reports. DOE/ET-0114/1. U.S. Department of Energy, Assistant Secretary for Energy Technology, Washington, D.C.

Eichers, Theodore R., Paul A. Andrilenas, and Thelma W. Anderson. 1978. Farmers' Use of Pesticides in 1976. Agricultural Economic Report No. 418. U.S. Department of Agriculture, Economics, Statistics, and Cooperatives Service, Washington, D.C.

Energy Research Advisory Board. 1980. Report on Gasohol. U.S. Department of Energy, Washington, D.C.

FEDS, 1981. Crop Production Budgets. Firm Enterprise Data System. U.S. Department of Agriculture, Economics, Statistics, and Cooperatives Service, Stillwater, Oklahoma.

Ferchak, John D., and E. Kendall Pye. 1981. Utilization of biomass in the U.S. for the production of ethanol fuel as a gasoline replacement--I. Solar Energy 26(1): 9-16.

Frederick, Kenneth D. 1980. Irrigation and the future of American agriculture. In Sandra S. Batie and Robert G. Healy (eds.), The Future of American Agriculture as a Strategic Resource, pp. 157-190. The Conservation Foundation, Washington, D.C.

Gormly, J. R., and R. F. Spalding. 1979. Sources and concentrations of nitrate-nitrogen in ground water of the Central Platte region, Nebraska. Ground Water 17(3): 291-301.

Grant, Kenneth E. 1975. Erosion in 1973-74: The record and the challenge. Journal of Soil and Water Conservation 30(1):29-32.

Hertzmark, Donald, Daryl Ray, and Gregory Parvin. 1980. The Agricultural Sector Impacts of Making Ethanol from Grain. SERI/TR-352-554. Solar Energy Research Institute, Golden, Colorado.

Hohmann, Michael A. 1980. On-Farm Production of Fuel-Alcohol in Mid-America--Technical and Economic Potential. MASEC-TP-80-009. Mid-America Solar Energy Complex, Minneapolis, Minnesota.

Laflen, J. M., and W. C. Moldenhauer. 1979. Soil and water losses from corn-soybean rotations. Soil Science Society of America Journal 43(6):1213-1215.

Lagrone, William Franklin. 1979. The Great Plains. In Another Revolution in U.S. Farming?, pp. 335-361. Agricultural Economic Report No. 441. U.S. Department of Agriculture, Economics, Statistics, and Cooperatives Service, Washington, D.C.

Lee, Linda K. 1978. A Perspective on Cropland Availability. Agricultural Economic Report No. 406. U.S. Department of Agriculture, Economics, Statistics, and Cooperatives Service, Washington, D.C.

Lee, Linda K. 1981. Cropland availability: The landowner factor. Journal of Soil and Water Conservation 36(3): 135-137.

Lockeretz, William, 1980. Grain Production for Alcohol Fuels. Contractor's report prepared for the U.S. National Alcohol Fuels Commission. Available from National Technical Information Service.

Lockeretz, William. 1978. The lessons of the Dust Bowl. American Scientist 66(5):560-569.

Meekhof, Ronald, Mohinder Gill, and Wallace Tyner. 1980. Gasohol: Prospects and Implications. Agricultural Economic Report No. 458. U.S. Department of Agriculture, Economics, Statistics, and Cooperatives Service, Washington, D.C.

Nebraska Department of Agriculture. 1979. Nebraska Agricultural Statistics, Annual Report, 1977-78. Nebraska Crop and Livestock Reporting Service, Lincoln.

OTA. 1979. Gasohol: A Technical Memorandum. Congress of the United States, Office of Technology Assessment, Washington, D.C.

OTA. 1980. Energy from Biological Processes. Volume II: Technical and Environmental Analyses. Congress of the United States, Office of Technology Assessment, Washington, D.C.

Sanghi, Ajay K., and David Johnson. 1978. The Ground Water and Energy Supply Situation for Great Plains Irrigation. CBNS-AE-10. Center for the Biology of Natural Systems, Washington University, St. Louis, Mo.

Schnittker Associates. 1980. Ethanol: Farm and Fuel Issues. Prepared for the U.S. National Alcohol Fuels Commission, Washington, D.C.

Spalding, R. F., J. R. Gormly, B. H. Curtiss, and M. E. Exner. 1978. Nonpoint nitrate contamination of ground water in Merrick County, Nebraska. Ground Water 16: 86-95.

Stewart, B. A., D. A. Woolhiser, W. H. Wischmeier, J. H. Caro, and M. H. Frere. 1976. Control of Water Pollution from Cropland. Volume II: An Overview. EPA-600/2-75-026b. U.S. Department of Agriculture, Agricultural Research Service, Washington, D.C.

Tyner, Wallace E. 1980. The Potential of Using Biomass for Energy in the United States. Publication Series No. 80-3. Energy Policy Research and Information Program, Purdue University, West Lafayette, Indiana.

USDA. 1979. Agricultural Statistics, 1979. U.S. Department of Agriculture, Washington, D.C.

USDA. 1980a. Soil and Water Resources Conservation Act Appraisal 1980, Review Draft, Part 1. U.S. Department of Agriculture, Washington, D.C.

USDA. 1980b. Report and Recommendations on Organic Farming. U.S. Department of Agriculture, Washington, D.C.

USDA. 1980c. Crop Production, 1979 Annual Summary. U.S. Department of Agriculture, Economics and Statistics Service, Washington, D.C.

USDA. 1981. Economic Indicators of the Farm Sector, Production and Efficiency Statistics, 1979. Statistical Bulletin No. 65. U.S. Department of Agriculture, Economics and Statistics Service, Washington, D.C.

U.S. National Alcohol Fuels Commission. 1981. Fuel Alcohol: An Energy Alternative for the 1980s. Final Report. Washington, D.C.

Welch, L. Fred. 1979. Nitrogen Use and Behavior in Crop Production. Bulletin 761. Agricultural Experiment Station, University of Illinois, Urbana-Champaign.

Wischmeier, W. H., and D. D. Smith. 1978. Predicting Rainfall Erosion Losses: A Guide to Conservation Planning. Agriculture Handbook No. 537. U.S. Department of Agriculture, Science and Education Administration, Washington, D.C.

Wisner, R. N., and J. O. Gidel. 1977. Economic Aspects of Using Grain Alcohol as a Motor Fuel, with Emphasis on By-Product Feed Markets. Economic Report Series, No. 9. Department of Economics, Iowa State University, Ames.

W. E. Larson, J. B. Swan,
F. J. Pierce

6. Agronomic Implications of Using Crop Residues for Energy

Crop residues, the above-ground plant tissue remaining in the field after harvest, are produced in large quantities yearly in the United States. Crop residues potentially can make a significant contribution to the nation's energy supply, either by their conversion to liquid or gaseous fuels, or by direct burning. However, not all residues should be regarded as available for removal because to varying degrees, maintaining soil productivity requires that crop residues be returned to the soil.

The flow chart in Figure 1 illustrates the important relationships between crop residues returned to the soil and various soil components and processes. Crop residues return plant nutrients, supply organic matter to soil, and influence soil temperature and soil water. These soil components in turn affect the rate at which residues decompose. Proper management of residues can greatly reduce soil and water losses from the land.

In this paper we discuss the amount of crop residues produced, the influence of these residues on soil properties, and the amount of residues available for removal assuming that erosion is the factor that most limits their removal.

Contribution from the Soil and Water Management Unit, North Central Region, Science and Education Administration, Agricultural Research, U.S. Department of Agriculture, St. Paul, MN, in cooperation with the Minnesota Agricultural Experiment Station, Paper No. 11640, Scientific Journal Series. Support in part by funds from the U.S. Department of Energy is gratefully acknowledged.

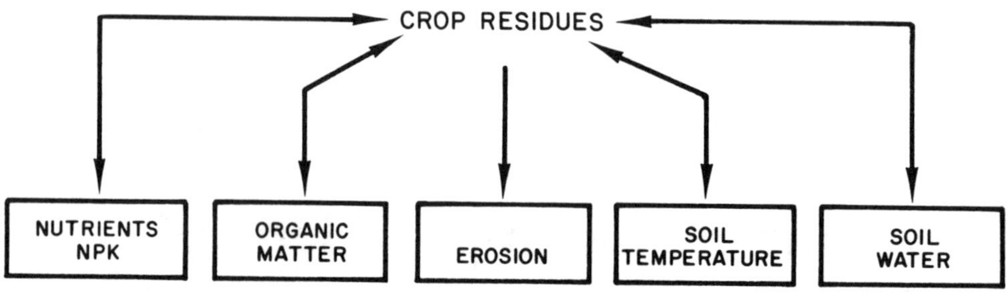

Figure 1. The relationship between crop residues and soil components and processes.

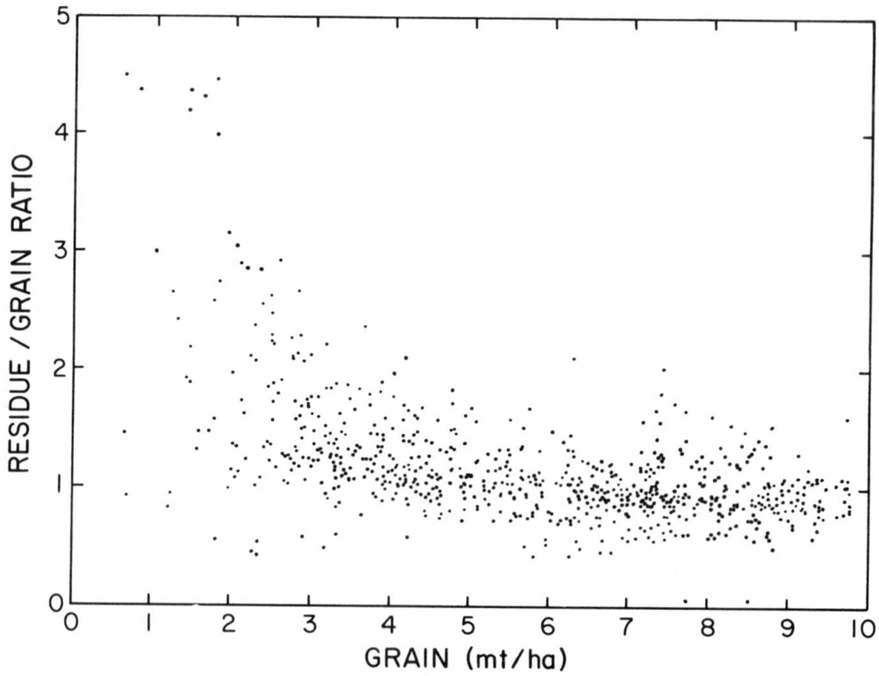

Figure 2. Scatterplot of the straw to grain ratio versus grain yield for corn.

Residue Production

Larson, Holt and Carlson (1978) estimated that 363 million metric tons (400 million tons) of crop residues are produced each year by the nine leading crops in the United States. This figure is based upon the total grain production for 1973 (U.S. Department of Agriculture, 1975) and the residue to grain ratios given in Table 1.

The quantity of crop residues produced is generally not measured, but rather is estimated by multiplying grain yield by the ratio of residue to grain. Residue/grain ratios for any crop vary considerably by year, location, crop cultivar, and soil fertility, with yearly weather changes accounting for the major portion of variability (Pierce and Larson, unpublished). The ratios decrease with grain yield, as illustrated in Figure 2 for corn. A decrease in the ratio for spring wheat was reported by Bauer and Zubriski (1978). Residue/grain ratios do not necessarily reflect harvestable residues. For example, in the case of soybeans, the leaves have fallen off the plant through senescence prior to seed harvest. In spite of variability, the mean ratios presented in Table 1 provide a reasonable estimate of national residue production.

Corn, wheat, and soybeans account for 75% of the total residues produced, with corn contributing about one-third more than wheat and twice as much as soybeans. Corn and soybean production are concentrated in the Corn Belt and small grain production (especially wheat) is concentrated in the Great Plains, making these two regions the largest residue producers in the United States. The combined residue production for corn, soybeans, and small grains for the Corn Belt and Great Plains are shown in Figures 3 and 4, respectively. Other areas (not shown) with concentrated residue production include: the Southeast Coastal Plain (corn, soybeans, cotton and rye); southern Mississippi Valley (rice, soybeans, and cotton); southeast Texas coast (sorghum and rice); Pacific Northwest (wheat and barley); and California (rice, corn, sorghum, wheat, barley, and oats).

Nutrients

Residues from the nine leading crops in all states contain about 4 million metric tons of nitrogen (N), 0.5 million metric tons of phosphorus (P), and 4 million metric tons of potassium (K), equal to about 40%, 10%, and 80% of the N, P, and K, respectively, in current fertilizer applications to all crops. The N, P, and K found in crop

Table 1. Ratio of residue (dry weight) to harvested portion (reported moisture basis) for selected crops.

Crop	Ratio (Residue/harvested portion)
Barley (Hordeum vulgare L.)	1.5
Corn (Zea mays L.)	1.0
Cotton (Gossypium hirsutum L.)	1.0
Oats (Avena sativa L.)	2.0
Rice (Oryza sativa L.)	1.5
Rye (Secale cereale L.)	1.5
Sorghum (Sorghum bicolor (L.) Moench)	1.0
Soybeans (Glycine max (L.) Merr.)	1.5
Wheat (Triticum aestivum L.)	
Winter	1.7
Spring	1.3

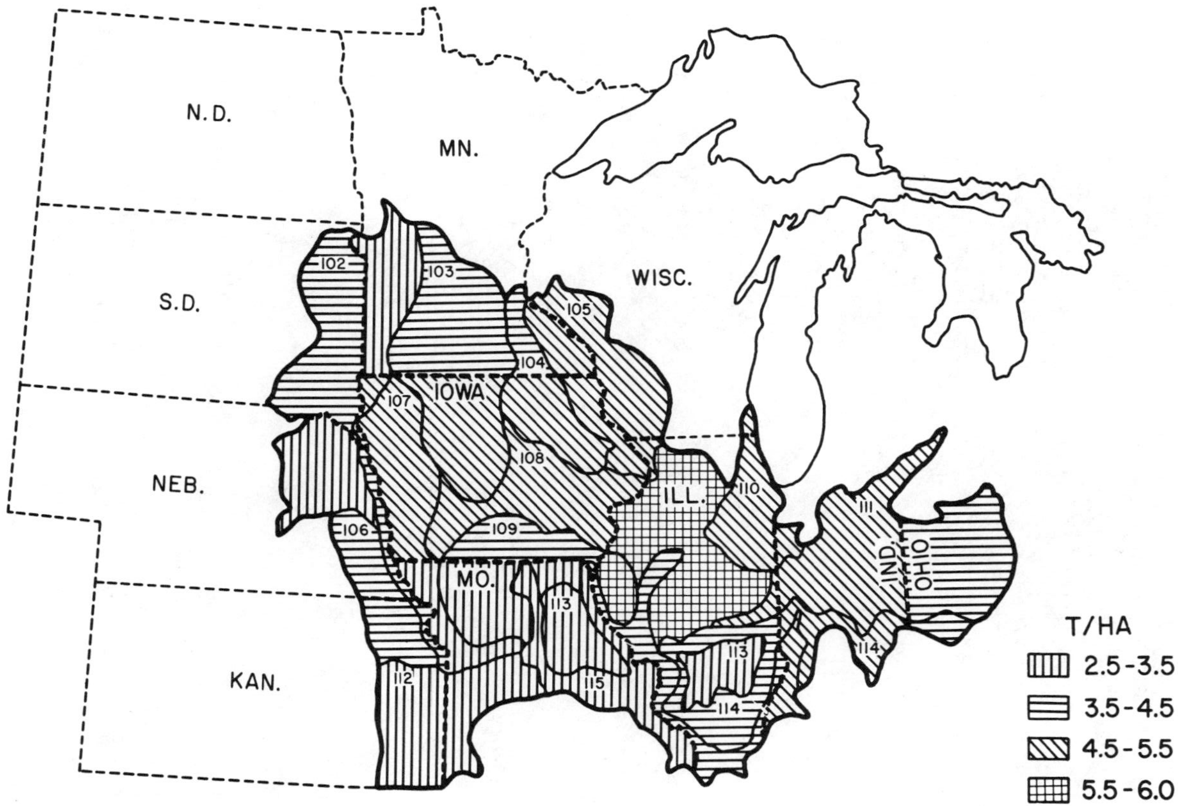

Figure 3. Combined residue production for corn, soybeans, and small grains in the Corn Belt (metric tons/hectare/year).

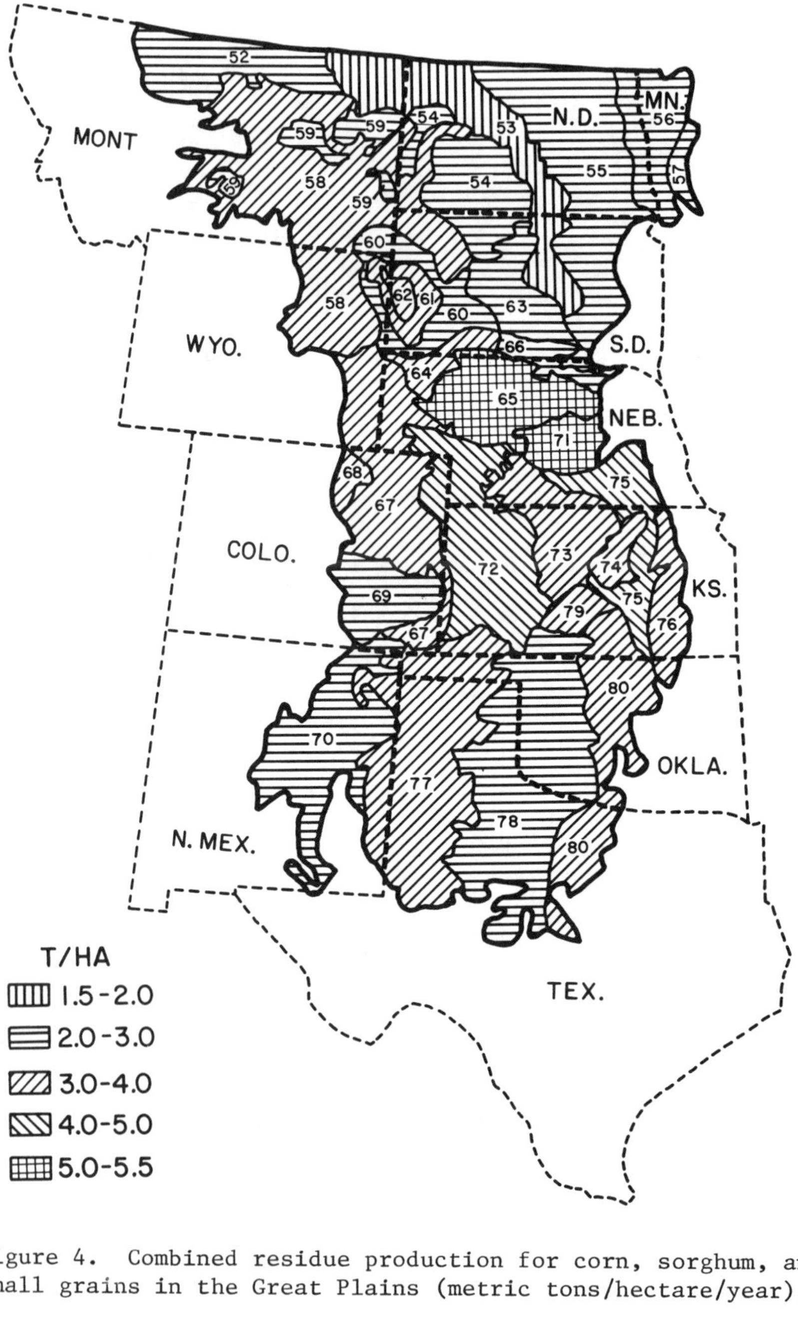

Figure 4. Combined residue production for corn, sorghum, and small grains in the Great Plains (metric tons/hectare/year).

residues for the 20 leading states, in absolute amounts and as a percentage of the commercial fertilizer sold for all crops, are given in Table 2. The total N, P, and K in plant residues was estimated by multiplying the quantity of residue produced by average values for residue compositon (Table 3). The levels of N and P are higher in grain than in crop residue, but the reverse is generally true for K.

Any system that removes residues must consider the return of the nutrients associated with the residue if long-term crop production levels are to be maintained. Nitrogen is particularly important because it is usually the most limiting element for nonleguminous plants and because its removal in crop residues can be substantial. For example, the N in crop residues in Iowa amounts to about 120 kg/ha (107 lb/A) for all the land planted to corn in the state. Phosphorus removal is appreciably lower than N removal because of the low P concentrations in the residues. Potassium levels in residue are relatively high, so that the total amount equals or exceeds the fertilizer applied in most midwestern and western states. While consideration must be given to nutrient removal in residues for most soils, some soils can continue to supply nutrients in a form available to plants for long periods of time. An example is the semi-arid and arid soils of the western U.S., which are generally well supplied with K.

Large amounts of nutrients are also contained in eroded sediments. Management of crop residues affects the amount of erosion and the quantity of nutrients in the sediment removed from a field. We estimated the average amount of N and P removed in eroded sediments for certain Major Land Resource Areas (MLRA) (Austin, 1965) in the Central and Southeastern states. These estimates were made by multiplying the estimated soil erosion rate (discussed later) by the average soil nutrient content and by an enrichment ratio. The last factor, which reflects the higher nutrient levels in the eroded fraction of the soil, was taken to have an average value of 2.0. (Barrows and Kilmer (1963) reported enrichment ratios for N, P, K, and organic matter in the range 1 to 5, with 2 perhaps the most common.)

The values given in Table 4 show that for fourteen MLRA's in the Corn Belt, the computed N removal was reduced from a mean of 64 kg/ha (57 lb/A), if all residues are removed, to 20 kg/ha by retaining 3920 kg/ha (1.5 tons/A) of crop residue on the soil surface. In MLRA 107 (Iowa and Missouri Deep Loess Hills, with rolling topography) the reduction due to residue retention would be from 178 to 64 kg/ha, while in MLRA 111 (Indiana and Ohio Till Plain,which is level to gently undulating) the reduction would be 29 to 8 kg/ha. For P, the

Table 2. Quantities of N, P, and K in crop residue, by weight and as percentages of commercial fertilizer use.

State	N		P		K	
	Thousand Metric Tons	% of comm. fert. N	Thousand Metric Tons	% of comm. fert. P	Thousand Metric Tons	% of comm. fert. K
Ark.	126	91	13	29	80	112
Calif.	51	10	7	10	74	139
Colo.	37	35	5	28	51	672
Ill.	553	84	72	38	482	106
Ind.	286	87	38	33	257	84
Iowa	597	88	82	52	556	160
Kans.	249	49	31	43	313	671
Mich.	63	44	9	15	70	52
Minn.	318	85	45	43	341	140
Mo.	199	62	24	32	155	95
Mont.	44	105	5	23	69	2,163
Nebr.	261	49	38	65	267	618
N. C.	70	35	9	15	62	40
N. Dak.	104	87	14	33	167	1,258
Ohio	162	62	21	20	143	61
Okla.	69	36	8	19	95	304
S. Dak.	94	100	15	63	140	1,153
Tex.	202	28	28	23	254	253
Wash.	32	17	3	13	47	235
Wis.	65	51	11	20	89	44
Total for 20 States	3,582		478		3,712	

Source: Larson, Holt, and Carlson (1978).

Table 3. Average N, P, and K concentrations in crop residues (%).

Crop	N	P	K
Barley	0.75	0.11	1.25
Corn	1.11	0.18	1.33
Cotton	1.75	0.22	1.45
Oats	0.63	0.16	1.65
Rice	0.60	0.09	1.16
Rye	0.50	0.12	0.70
Sorghum	1.08	0.15	1.32
Soybeans	2.25	0.22	1.05
Wheat	0.67	0.07	0.97

Source: The Fertilizer Institute (1962).

Table 4. Estimated average amounts of N and total P in eroded sediments from two residue management systems in fourteen major land resource areas in the Corn Belt.

MLRA[1]	Nitrogen			Phosphorus		
	In soil (%)	In eroded sediment (kg/ha)		In soil (%)	In eroded sediment (kg/ha)	
		Residues removed	Residues returned		Residues removed	Residues returned
102	0.2	57	22	0.05	14	6
103	0.2	39	16	0.05	10	4
104	0.2	54	22	0.05	13	5
105	0.2	95	33	0.05	24	8
106	0.1	57	20	0.05	28	10
107	0.2	178	64	0.05	44	16
108	0.1	55	17	0.05	28	9
109	0.1	69	19	0.05	34	10
110	0.2	66	17	0.05	16	4
111	0.1	29	8	0.05	14	4
112	0.1	35	9	0.05	17	4
113	0.1	59	13	0.05	29	6
114	0.1	51	13	0.05	25	6
115	0.08	50	12	0.05	31	8
Mean		64	20		23	7

[1] Major Land Resources Area (Austin, 1965). See Figure 3.

reductions are similar but the total amounts involved are smaller.

Soil Organic Matter

There is an ancient saying that "corruption is the mother of vegetation". Its basis, according to Russell (1973), was the early discovery that organic materials, manures, composts, and dead animal bodies all increased the fertility of the land. Thus, organic matter and its maintenance has long been associated with productivity of the soil.

Organic matter influences the physical and chemical properties of soils far out of proportion to the small quantities present (Brady, 1974). It is a key property in soil water balance, soil tilth, soil erodibility, and nutrient cycling and availability to plants (Cooke, 1977). Changes in soil organic matter, therefore, can have profound effects on the soil system.

Under natural conditions, soil organic matter is in equilibrium with the soil forming factors. In order of their importance to organic matter content of soils, these factors are climate, vegetation, topography, parent material, and age (Jenny, 1930). Cultivation and cropping of soils has seriously altered this equilibrium, resulting in significant reductions in the organic matter content of soils (Campbell, Paul, and McGill, 1976; Haas, Evans, and Miles, 1957; Jenkinson and Rayner, 1971; Jenny, 1941; Paul and Van Veen, 1978; and Stevenson, 1965). As illustrated in Figure 5, the rate of change of organic matter is highest in the first 10 to 20 years and organic levels are thought to stabilize some 50 to 100 years after cultivation commences (Jenny, 1941). This new equilibrium level will be reached where cropping practices have continued uninterrupted for a long period of time (Barber, 1979; Campbell, Paul, and McGill, 1976).

In the Corn Belt, the return of crop residues is considered necessary to maintain soil organic matter at its present level (Barber, 1979). However, modification of cropping practices will result in a new equilibrium. In a study conducted on a silt loam soil in southwest Iowa, Larson et al. (1972) reported that after 11 years of continuous corn, the organic carbon content was a linear function of the amount of residue added and was independent of the type of residue added (Figure 6). They estimated that about 5 metric tons/ha/yr of residue, or the amount produced by a 5 metric ton/ha corn crop, was needed to maintain the initial 1.8% carbon level in the soil. (To put the organic carbon data of Figure 6 on an approximately equivalent basis with the organic

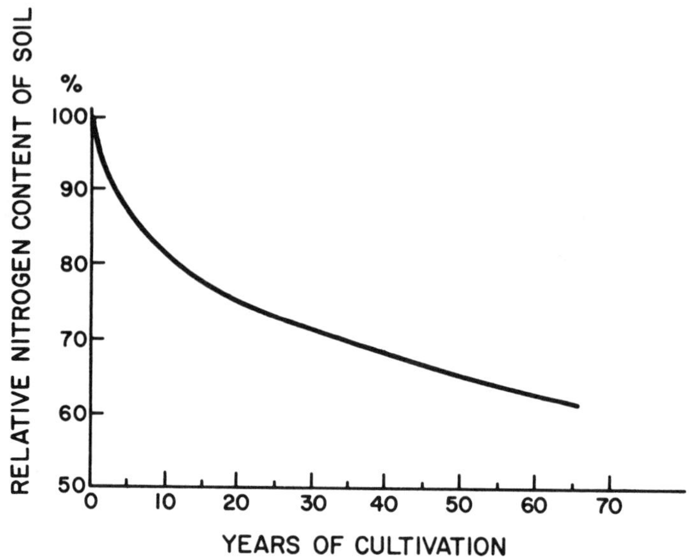

Figure 5. Decline of soil nitrogen under average farming conditions in the Central States (adapted from Jenny, 1941).

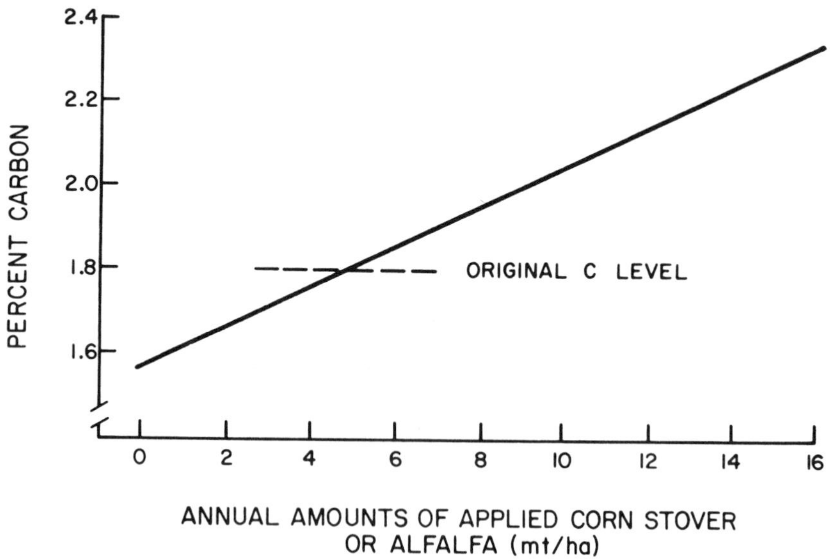

Figure 6. Carbon content of a silt loam soil in southwest Iowa as influenced by amounts of corn stover or alfalfa hay added to the soil for each of 11 consecutive years. Soil was cropped to corn with conventional tillage (adapted from Larson et al., 1972).

matter data used in some papers, multiply by 1.72.) In the Pacific Northwest, Rasmussen et al. (1980) reported that yearly additions of various organic materials, even at high rates, were insufficient to maintain soil organic matter over a 45-year period. They did report, however, that the change in carbon was linearly related to the amount of carbon added to the soil, which is in agreement with the findings of Larson et al. (1972).

The question can be raised as to how much crop residue removal would alter soil organic matter levels, and hence long-term soil productivity (Barber, 1979). The amount and type of tillage and amount of residue that tillage incorporates into the soil affect the rates of residue decomposition and mineralization. Tillage practices which minimize soil disturbance and retain all or part of the residues on the soil surface reduce these rates significantly. Under such conditions, it may be possible to remove part of the residues and still maintain soil organic matter. For example, under 6 years of no-tillage in Minnesota with all residue retained on the surface, crop residues accumulated to the extent that a mat of residues formed at the soil surface (Personal communication, G.W. Randall).

The effects of crop residue removal on soil organic matter should be known before the decision to remove residues for alternate uses is made. This requires an understanding of the decomposition process and the factors that affect it, including the effects of crop residue management and tillage practices.

Soil Erosion

Soil erosion can reduce the productive capacity of soil by removing important quantities of plant nutrients and organic matter, by reducing the soil's water holding capacity, by reducing the effective plant rooting depth, by reducing infiltration rates, and by destroying desirable soil structure. The degree of damage to these soil properties depends upon the amount of erosion and on the properties of both the surface soil and subsoil and their depth. If the soil material is deep, with favorable characteristics throughout its total depth, and if plant nutrients are supplied with fertilizer, erosion may not reduce the productive capacity unless it continues for a long time. Examples of such soils are the deep, permeable, loess-derived soils of the Missouri and Mississippi river basins. On the other hand, if the surface soil consists of only a few feet of desirable material, which is underlain by an undesirable material, erosion losses may be serious from the start.

Undesirable subsoil materials include rock, unconsolidated materials, or materials high in sand or clay or highly acid.

In a medium-textured Corn Belt soil, water available to plants will amount to a maximum of about 16% of the soil volume. Thus, if a permeable soil is underlain by consolidated materials, and if 1.0 cm of surface soil is removed by erosion, then 0.16 cm of available water storage capacity will be lost. Likewise, an equal amount of rooting depth for sorption of nutrients and water is lost. Leeper, Runge, and Walker (1974) found that rooting depth and available water holding capacity were closely correlated with corn yield (R^2 values of 71% and 73%) in studies conducted in Illinois. They concluded that with adequate fertility and a high level of management, the potential of a soil to produce corn was largely determined by the soil's capacity to store and supply water. Loss of 1 cm of rooting depth reduced yields an average of 52 kg/ha (2.1 bu/A per inch of rooting depth), while loss of 1 cm of available water holding capacity caused an average yield reduction of 173 kg/ha. However, losses varied considerably with growing conditions, ranging from 7 to 89 kg/ha/cm of rooting depth and from 25 to 339 kg/ha/cm of available water holding capacity.

Removal of relatively high organic matter surface soil and exposure of low organic matter subsoil will often reduce the tilth of the soil. Poor soil tilth may result in cloddy soil, an increased need for tillage operations, and uneven plant seedling emergence.

Lindstrom et al. (1979), using procedures outlined by Gupta, Onstad, and Larson (1979), estimated the effect of residue management and tillage on soil erosion by water with the Universal Soil Loss Equation (USLE). The estimated soil loss depends on rainfall amount and intensity, soil erodibility, the length and gradient of the slope, crop management, and erosion control practices (Wischmeier and Smith, 1978).

The USLE was used with basic soil and landscape data by MLRA from the SCS Conservation Needs Inventory (USDA, 1971). An erosion control practice factor, P, of 1.0 was used, which corresponds to up-and-down slope cultivation. Judgement was used to fit commonly used rotations to each soil mapping unit (combination of soil type/slope/previous erosion) until the weighted percentage of crops agreed with the tabulated average values for an MLRA from the Statistical Reporting Service (USDA, 1975). The crop management factor was then calculated for each crop rotation and several tillage-management practices, with the calculated erosion rates for each soil mapping unit then combined to give a weighted average for each MLRA.

The estimated erosion rate for conventional tillage, where all residues are removed, and for conservation and no-till systems, where 3920 kg/ha of crop residues are left on the soil surface, are given in Table 5. Conventional tillage consisted of fall plowing, spring disking, and harrowing. Conservation tillage was equivalent to a subsurface tillage system which leaves 66% of the surface covered by residue. In the no-till system, 90% surface coverage was assumed. Small grain and soybean residue was assumed to be twice as effective for erosion control as corn residue on a weight basis. The soil tolerance value (T) (Table 5) is the maximum soil loss consistent with sustained productivity. The data show that the mean erosion rate would exceed T on 13 of the 14 MLRA's if all residues were removed, whereas the mean erosion rate would exceed T on only 3 of the 14 MLRA's if 3920 kg/ha were left on the soil surface in a no-till system.

Using procedures similar to those of Gupta, Onstad, and Larson (1979), Campbell et al. (1979) showed that in the Southeast, the weighted average soil loss with conventional tillage and residues removed exceeded tolerance limits in 13 of 14 MLRA's studied. Conservation tillage significantly reduced potential soil losses in all MLRA's. In four of six states, 60% of the crop residues were needed for water erosion control. About 90% of the residues were needed for water erosion control in Alabama and Mississippi. Allmaras et al. (1979) found that in the Northwest wheat and range region of Oregon, 60% of the residues could be harvested from 88% of the small grain area.

Skidmore, Kumar, and Larson (1979), using the Wind Erosion Equation, found that enough residues are produced to protect against wind erosion over about 55% of the Great Plains, assuming the soil is smooth and the residues are left on the surface.

Soil Water

Crop residues play an important role in the water cycle in soil. Their presence on the soil surface modifies the soil-air interface, which greatly affects the soil water cycle.

The cycle of water in soil consists of distinct stages or processes--runoff, infiltration, redistribution, evaporation

Table 5. Estimated mean soil water erosion in the Corn Belt with and without crop residues on the soil surface and average tolerance (T) [metric tons/ha/yr].

MLRA[a]	States	Conventional tillage All residues removed	Conservation tillage 3920 kg/ha residues on soil surface	No-till 3920 kg/ha residues on soil surface	T
102	SD, NE, MN, IA	14.2	6.9	5.6	10.9
103	MN, IA	9.7	4.5	4.0	10.9
104	MN, IA	13.4	6.6	5.4	10.4
105	WI, MN, IA, IL	23.7	10.7	8.2	10.2
106	NE, KS	28.5	11.8	10.1	9.4
107	IA, MO	44.7	19.2	16.0	10.9
108	IA, IL	27.6	11.4	8.7	10.6
109	IA, MO, IL	34.6	13.0	9.7	9.1
110	IL, IN	16.4	6.3	4.3	9.8
111	IN, OH, IL	14.4	6.2	4.0	9.3
112	KS, MO	17.4	6.3	4.5	9.0
113	MO, IL	29.4	10.2	6.6	8.0
114	IL, IN	25.3	9.4	6.4	8.8
115	MO, IL, IN	31.4	11.9	7.8	10.1

Source: Adapted from Lindstrom et al. (1978).

(a) See Figure 3 for locations.

and transpiration, percolation, and storage--that are interdependent and which may occur simultaneously.

Precipitation is partitioned into interception by vegetation (living or dead), infiltration and runoff. Infiltrated water undergoes redistribution and storage as modified by evaporation and transpiration, and can be accompanied by deep percolation out of some arbitrarily defined rooting volume. Upward water movement into the rooting volume can also occur.

Each process in the water cycle proceeds at a rate determined by the interaction of climate, soil, and biota, and through these by crop residue management and tillage practices. Climate factors include rainfall amount, intensity, and frequency, net radiation, vapor pressure, wind, and temperature. Soil factors include water retention and flow characteristics, soil thermal properties, topography, and surface configuration. Biotic factors include crop canopy effects on soil microclimate and consumptive water use. Crop residues on the soil surface directly influence interception, runoff, infiltration, and evaporation, and indirectly influence other processes in the cycle, including water storage.

Interception

Crop residues can contain nearly four times their weight in water at saturation (Parr and Papendick, 1978). Thus, they act to reduce the amount of water available for infiltration and runoff by intercepting rainfall that is absorbed into the tissue and subsequently evaporated. Surface residues also increase the surface retention capacity of a soil and slow runoff, thereby increasing the amount of water available for infiltration and evaporation and further reducing runoff.

Runoff and Infiltration

By intercepting the impact energy of rainfall and increasing the water stability of soil aggregates, residues inhibit the sealing or crusting process caused by the beating action of raindrops or slaking of the soil. Such sealing is particularly likely to occur on soils with unstable structure (Van Doren and Allmaras, 1978), but it also can occur on many relatively stable medium- and fine-textured soils if sufficient raindrop energy is applied (Burwell and Larson, 1969). Van Doren and Allmaras (1978) showed that the saturated hydraulic conductivity of a given soil decreases exponentially as rainfall kinetic energy increases. This reduction increases the potential for runoff. Conversely, residues reduce potential runoff by reducing rainfall kinetic energy.

Evaporation

Mulches of crop residue can slow the evaporation of water from the soil by acting as a heat barrier, by reducing the transport of water vapor from the soil surface, and in some cases by reflecting solar radiation, which reduces the amount of energy for evaporation (Massee and Cary, 1978). The effectiveness of such a mulch in decreasing evaporation also depends greatly on the relative wetness of the soil surface (Massee and Cary, 1978) and thus on the amount and frequency of rainfall. Bond and Willis (1969) showed that cumulative evaporation decreased as wheat straw mulch rate increased, but the cumulative evaporation became nearly the same for all mulch rates when evaporation continued without additional rainfall for a sufficiently long time.

Storage

Increased soil water storage may be due to either increased infiltration, decreased evaporation or both. Research studies on many different soils have shown that residue management through conservation tillage can increase stored water by reducing runoff.

In general, crop residues on the soil surface will reduce runoff from storms of all sizes and eliminate runoff from small storms. The amount by which surface residues increase the total retained soil water depends on the infiltration rate of the soil (Onstad and Otterby, 1979). For those soils which have moderate infiltration rates when thoroughly wetted, the increase in retained water ranges from 5 mm in the Great Plains to 50 mm in the Southeast United States. For soils with very slow infiltration rates when thoroughly wetted, the increase in retained water ranges from 25 mm in the Great Plains to 125 mm in the Southeast. Some portion of the retained water may be subject to drainage or deep percolation depending on the circumstances.

These changes can significantly affect crop yields. For example, Greb (1979) estimates that 45% of the wheat yield increase in the Central Great Plains over the past 40 to 50 years was due to improved storage of water in fallow as a result of improved weed control and increased use of stubble mulch tillage (which leaves residues on the surface). Thus large scale removal of crop residues in the semi-arid Great

Plains regions could reduce current fallow efficiency. It would also restrict possible future increases in the water storage efficiency on land that now is fallowed with residues turned under (black fallow).[1]

The increases in water storage efficiencies of fallow with improved tillage systems and the use of crop residues can be dramatic. For example, Greb, Smika, and Welch (1979) state that "currently, fallow efficiency with no-till averages about 50% compared with about 20% for the original dust mulch system, and 24% for conventional black fallow." A straw mulch gave an average increase in soil water of 2.6 cm at the end of fallow compared to bare fallow at seven central Great Plains locations (Table 6; Greb, 1979). At four locations increased soil water storage at the end of fallow was directly related to the rate of straw mulch. A change of 2.5 cm of stored water corresponds to a yield change of 270 to 470 kg/ha (4 to 7 bu/A) for wheat (Greb, Smika, and Welsh, 1979).

Another special situation exists on Midwestern soils with unstable structure, where non-plowed tillage methods generally require high levels of residue cover to achieve their full potential for water conservation and yield. Van Doren (1965) determined that on soils that crusted readily, corn yields were directly proportional to the fraction of the soil surface covered with mulch for most nonplowed treatments tested. Reducing mulch levels under these conditions could significantly reduce corn production, as illustrated in Figure 7.

Snow Retention

Snow catchment is an important overwinter water source in the Central and Northern Great Plains. Allowing wheat stubble to stand over the winter months effectively increased the amount of winter precipitation stored in studies at several locations in the Central and Northern Great Plains (Smika and Whitfield, 1965). Where snow is retained, between 35 to 55% of water recharge during fallow may be from snow melt in stubble (Greb, 1979). Improved retention of snow in standing stubble could reduce the necessity of fallowing and might allow the winter wheat producing areas to move farther into the spring wheat area (Papendick, Swan, and Robocker, 1980).

---

1 "Fallow" refers to the practice of leaving land uncropped and weed free for periods of time to accumulate and retain water and mineralized nutrient elements.

Table 6. Additional increase in stored water during fallow due to straw mulch compared to bare fallow.

Locations (No. of Sites)[a]	Soil Water Increase (cm)
Colorado	3.1
Nebraska (2)	0.2, 5.7
Wyoming	1.4
Kansas (3)	0.4, 2.5, 4.9
Avg. of 7 sites	2.6[b]

Source: Adapted from Greb, Smika, and Welsh (1979).
[a]Total of 38 years of observation at all sites.
[b]Average increase due to bare fallow: 9.0 cm; mulch: 11.6 cm.

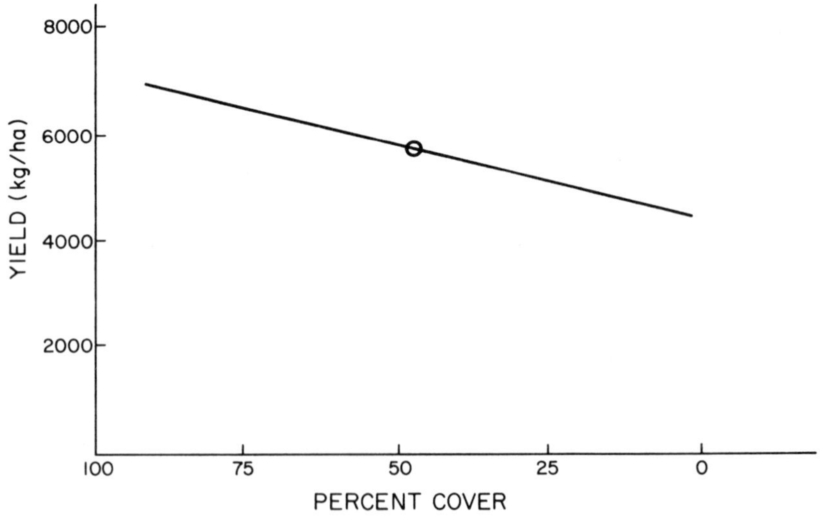

Figure 7. The effect of soil cover, or residue left on the ground after harvesting of the preceding crop, in no-tillage agriculture on the yield of corn on Wooster silt loam. The 0 shows the yield of conventional tillage: 5700 kg/ha. (Adapted with permission from Triplett and Van Doren, 1977.)

Soil Temperature

While crop residue management can have a material influence on soil temperature, crop residue removal is usually not limited by soil temperature considerations per se. The following discussion describes the effect of soil temperature and thus discusses the effect of residue induced soil temperature changes on crop growth and yield.

Modifications of the surface energy balance by crop residues affects both soil temperature and soil water. Residues reduce the daily and seasonal variations of temperature at a given depth and cause a lag in the seasonal soil temperature curves (Fig. 8). Thus, in the northern hemisphere in mid-latitudes, a mulched soil tends to warm more slowly than an unmulched soil in the spring and cool more slowly in the fall, and to be cooler in the spring and summer, but warmer in winter. In the spring and summer, residues tend to reduce daily average temperature and especially to reduce daily maximum temperature (Fig.9). The effect of residues on soil temperature decreases as a crop canopy progressively covers the soil surface.

Crop residue mulches influence the temperature of the soil in three ways: 1) by acting as an insulating layer on the soil surface, reducing the amount of heat that enters the soil; 2) by increasing the reflection coefficient of incident radiant energy (fresh crop residues are usually lighter colored than wet soil); and 3) by reducing soil water evaporation. Effect (3) tends to partially conteract the first two since a smaller fraction of the total heat generated at the soil surface is used as latent heat of vaporization. Thus, on a mulched soil, sensible heat constitutes a larger fraction of the total heat generated compared with the unmulched soil. When the soil is wetter due to the influence of a mulch, the soil has a greater specific heat capacity, which also slows the rate of change of soil temperature. Soil temperature and evaporation are also inter-related since a lower soil temperature decreases the vapor pressure of soil water, decreases the vapor pressure gradient, and thus the potential evaporation (Van Doren and Allmaras, 1978).

The temperature of the soil has an important effect on the growth of plants. Plants vary, however, in the cardinal (minimum, optimum and maximum) soil temperatures (Richards, Hagan, and McCalla, 1952; Willis and Amemiya, 1973). The cardinal temperatures also vary with stage of plant growth. Allmaras, Burrows, and Larson (1964) showed that the optimum soil temperature (measured at the 10 cm depth) for corn was approximately 27.4 degrees C during the first 6 weeks

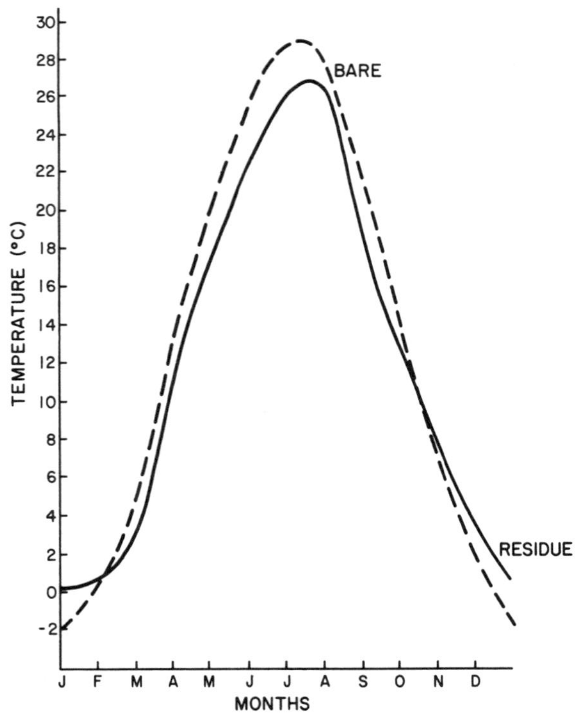

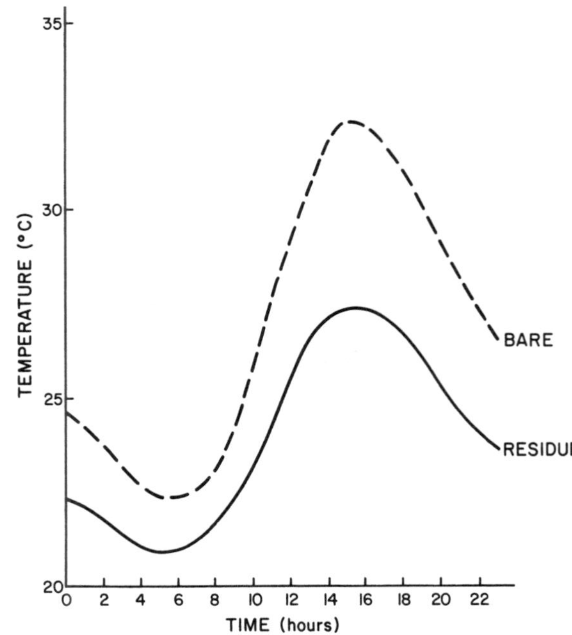

Figure 8. *(left)* Mean monthly soil temperature at the 2.5 cm depth for bare and residue covered (3.4 metric tons/ha wheat straw) surfaces at Lafayette, Indiana. Figure 9. *(right)* Soil temperature at the 5 cm depth during June at Morris, Minnesota, for a bare and residue covered (7 metric tons/ha oat straw) soil.

following planting (Fig. 10). During this period, the growing point of corn is below the soil surface, and thus is affected by the soil temperatures. At this time, soil temperature is generally well below the optimum for corn growth in the Northern Corn Belt.

Burrows and Larson (1962) have shown that if all the crop residues produced from an average 6000 kg/ha (96 bu/A) crop of corn were left on the surface the temperature of the soil at the 10 cm depth will be reduced by about 1.2 C. If the 10 cm temperature of a bare soil of 19 C (which is reasonable for Minnesota conditions) is reduced by 1.2 C, the early growth of corn would be reduced about 40% (calculated from Fig. 10). However, if a 10 cm depth bare soil temperature of 26 C (e.g. South Carolina) is reduced by 1.2 C, the early season corn growth could be reduced only by about 10 percent.

Thus, for field crops, beneficial temperature effects due to mulch are not common in northern climates, where soil temperatures are near the minimum for growth early in the season. However, in southern regions where bare soil temperatures are near the optimum, a change in a few degrees may have little influence on growth, while if the bare soil temperature is above the optimum, a reduction of a few degrees may increase plant growth.

In the tropics the combination of high soil temperature and low availability of soil water can adversely effect the yield of crops such as corn and soybeans, so that residue mulches may decrease soil temperatures and improve crop growth (Lal, DeVleeschawer, and Nganje, 1980).

Crop Residues Available for Removal

Crop residues should be considered available for removal for energy production only when they exceed the amount needed to maintain soil productivity. The problem, at present, is documenting the residue requirements for a given crop-soil-climate situation for the various soil components and processes previously discussed.

Recently, the amount of crop residues available for removal from the land were estimated based upon the assumption that residues were available for removal only if erosion rates did not exceed the soil tolerance limit. From the same data base used in those estimations, we compiled density maps which indicate the amount of residues produced in the Corn Belt and in the Great Plains and the fraction of residues available in those areas as well as in a six state

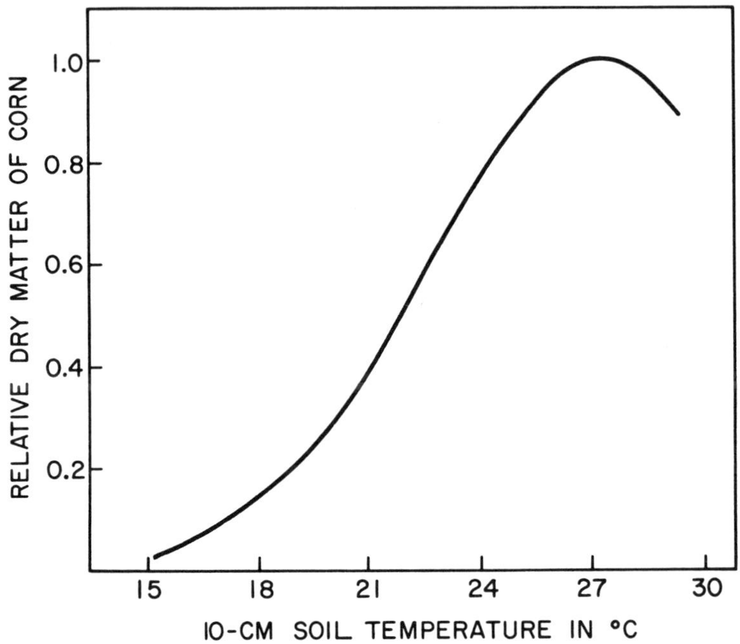

Figure 10. Relative yield of dry matter in young corn plants in the field as affected by average 10-cm soil temperature (from Allmaras, Burrows, and Larson, 1964).

area of Southeastern United States. The calculation procedure has been outlined by Gupta, Onstad, and Larson (1979) for water erosion and by Skidmore, Kumar, and Larson (1979) for wind erosion. Only wind erosion was considered in the Great Plains and only water erosion in the Corn Belt and Southeastern United States (Campbell et al, 1979). All water erosion amounts are estimates of soil removal from a land unit in the field, not sediment delivery to a stream.

The percentage of residues available for removal from the land in the Corn Belt are illustrated in Figure 11. These averages were calculated for each MLRA within each state in the Corn Belt (Lindstrom et al., 1979). The total residue production in the Corn Belt is about 141 million metric tons, of which 63% is from corn, 23% from soybeans, and 14% from small grains. Under conventional tillage, residues can be removed from about 36% of the cultivated land area in the Corn Belt without causing excessive water erosion. This area is increased to 78% when 3920 kg/ha of crop residues are left on the land and no-till management is used. In general, annual residue production is greatest in the central Corn Belt, but the greatest percentage available for removal is in the northwest Corn Belt because of gentler slopes and fewer highly erosive rainstorms. The areas with greatest availability of residues are also the areas having the greatest fraction of cultivated land in corn and where corn yields are higher than the national average.

The percentage of residues available in the Great Plains is illustrated in Figure 12 (Skidmore, Kumar, and Larson, 1979). Residue production in the Great Plains is about 75 million metric tons, of which 85% is wheat and corn residues. Of this, about 21% is available for removal if only wind erosion is considered. In general, residue production and availability are greatest in the more humid eastern section of the Great Plains.

Of the approximately 10 million metric tons of residues produced in the Southeastern United States, 60% are needed for water erosion control and 40% are available for removal (Fig. 13) (Campbell et al., 1979). In Alabama and Mississippi, less than 10% of the residues are available for removal; conversely, in the Atlantic Coastal Flatwoods region of Georgia (MLRA 153), 75% are available.

Of the approximately 1.34 million metric tons of residues produced in three MLRA's in eastern Oregon, about 60% could be removed without excessive water erosion on 88% of the harvested area (Allmaras et al., 1979). Use of a

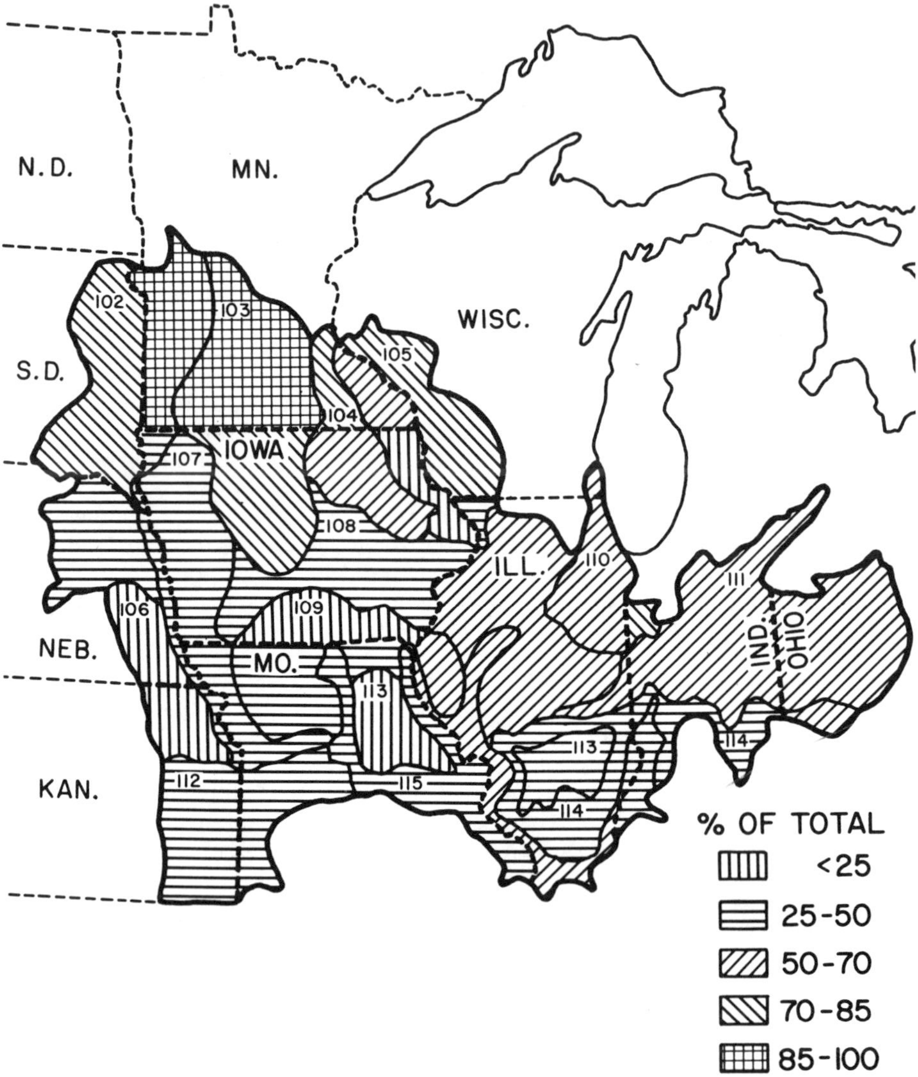

Figure 11. Percentage of total residues produced in the Corn Belt that are available for removal for energy or other uses.

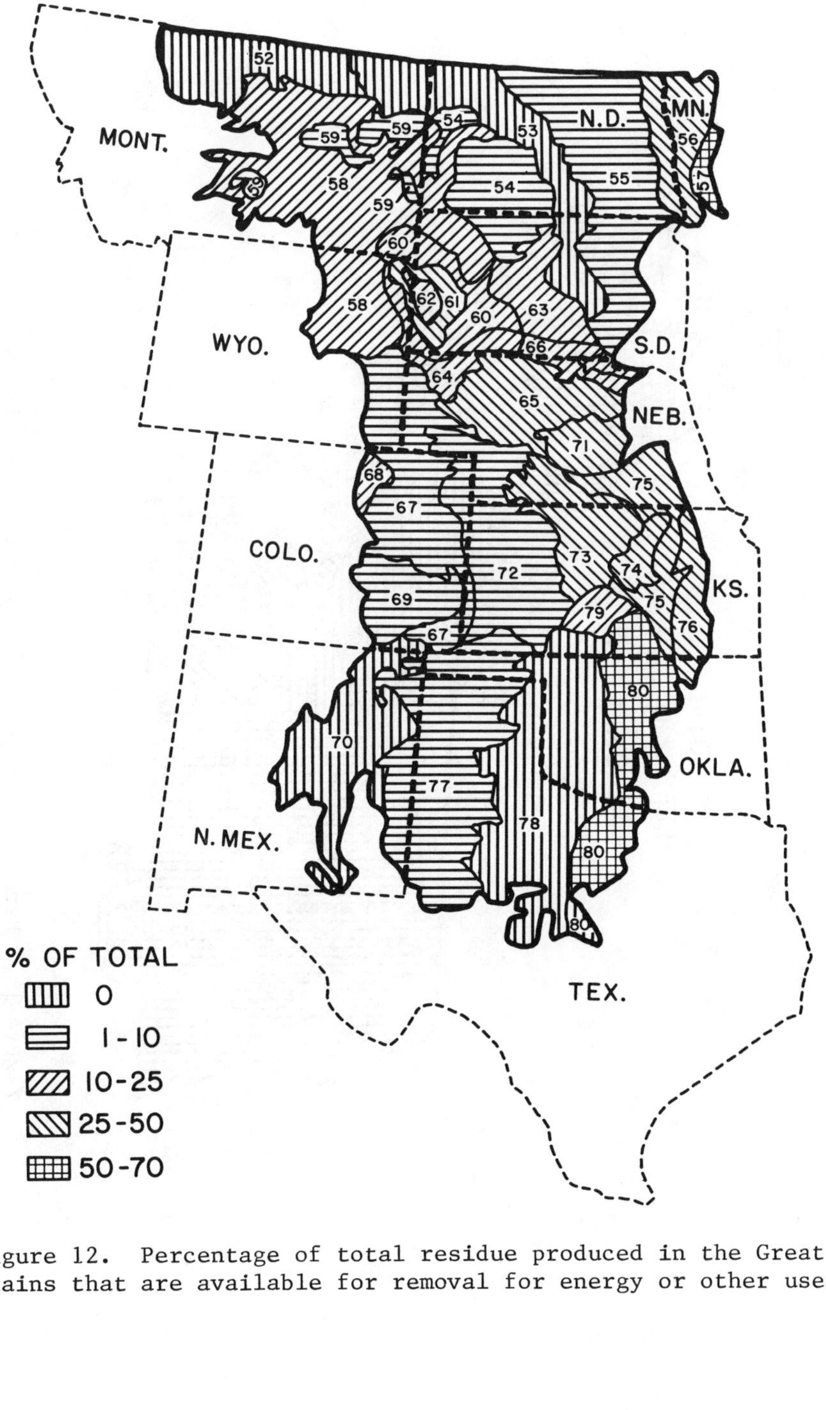

Figure 12. Percentage of total residue produced in the Great Plains that are available for removal for energy or other uses.

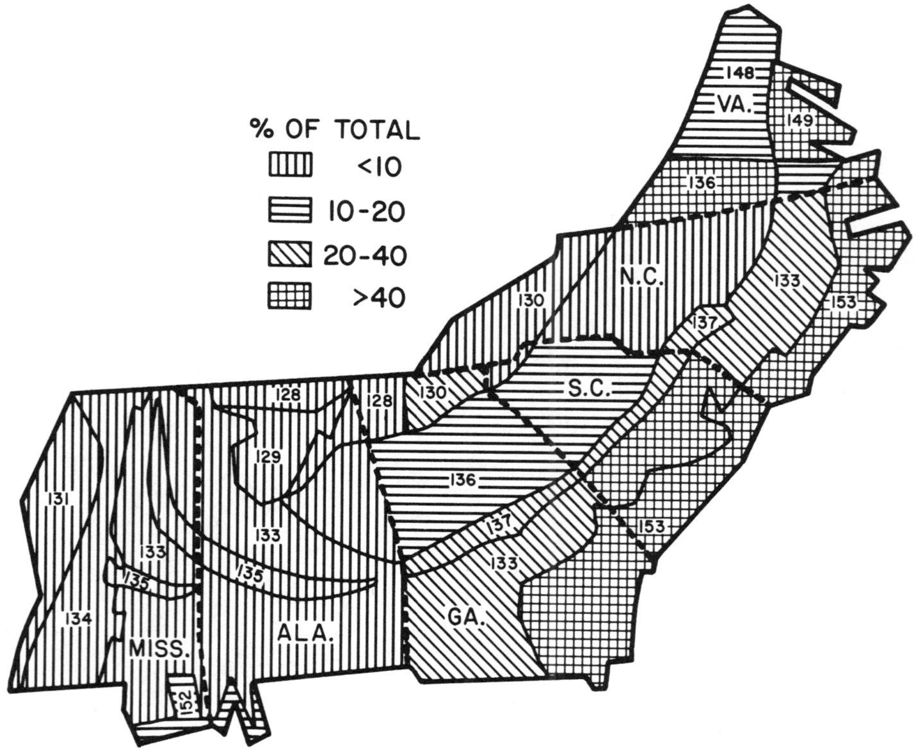

Figure 13. Percentage of total residue produced in the Southeast that are available for removal for energy or other uses.

reduced tillage system instead of conventional tillage would increase residue availability by about 20%.

These estimates of available residues for energy and other uses are based solely on soil erosion as the factor limiting soil productivity. However, while residue requirements for other soil components and processes are not well documented, it appears they may require more residues than are necessary for erosion control under many conditions. This is particularly true in dryland farming areas in the Great Plains, where the potential for increasing the efficiency of storage in precipitation would be severely restricted if residues were unavailable for water conservation. In some cases, residue removal would cause a reduction in the efficiency of storage below current levels. Substantial losses of soil organic matter and N have occurred as a result of cultivation. Possible further reductions in organic matter from residue removal should receive serious attention. Also, residue removal would require corresponding higher fertilizer applications in most cases. As long as agriculture can tolerate increased costs and while fertilizer supplies remain plentiful, nutrient removal should not be a factor limiting residue removal. Soil temperature is not usually a factor limiting residue removal in the U.S. because the effects of residues on soil temperature are beneficial only in a few specific situations.

This analysis assumes that current production practices are used. Improved conservation practices, growing of cover crops, and alternate crop sequences might increase the amount of crop residues available for removal. If the soil's needs can be met, however, with the partial or nearly complete removal of crop residues, there should be no objection by the agricultural sector.

References

Allmaras, R. R., S. C. Gupta, J. L. Pikul, Jr., and C. E. Johnson. 1979. Tillage and plant residue management for water erosion control on agricultural land in eastern Oregon. J. Soil Water Conserv. 34:74-76.

Allmaras, R. R., W. C. Burrows, and W. E. Larson. 1964. Early growth of corn as affected by soil temperature. Soil Sci. Soc. Am. Proc. 28:271-275.

Austin, M. E. 1965. Land resource regions and major land resource areas of the United States. Agric. Handb. No. 296, USDA. U. S. Government Printing Office, Washington, D. C.

Barber, S. A. 1979. Corn residue management and soil organic matter. Agron. J. 71:625-627.

Barrows, H. L. and V. J. Kilmer. 1963. Plant nutrient losses from soils by water erosion. Adv. Agron. 15:303-316.

Bauer, A. and J. C. Zubriski. 1978. Hard red spring wheat straw yields in relation to grain yields. Soil Sci. Soc. Am. J. 42:777-781.

Bond, J. J. and W. O. Willis. 1969. Soil water evaporation: Surface residue rate and placement effects. Soil Sci. Soc. Am. Proc. 33:445-448.

Brady, N. C. 1974. The nature and properties of soils. Macmillan, New York.

Burrows, W. C. and W. E. Larson. 1962. Effect of amount of mulch on soil temperature and early growth of corn. Agron. J. 54:19-23.

Burwell, R. E. and W. E. Larson. 1969. Infiltration as influenced by tillage induced random roughness and pore space. Soil Sci. Soc. Am. Proc. 33:449-452.

Campbell, C. A., E. A. Paul, and W. B. McGill. 1976. Effect of cultivation and cropping on the amounts and forms of nitrogen in soil. pp. 9-101. In W. A. Rice (Conf. Chm.) Proc. of Western Canada Nitrogen Symposium. Alberta Agriculture. Edmonton.

Campbell, R. B., T. A. Matheny, P. G. Hunt, and S. C. Gupta. 1979. Crop residue requirements for water erosion control in six southern states. J. Soil Water Conserv. 34:83-85.

Cooke, G. W. 1977. The roles of organic manures and organic matter in managing our soils for higher crop yields - a review of the experimental evidence. In Proc. of the Int. Seminar on Soil Environment and Fertility Management in Intensive Agriculture (Tokyo, Japan). SEFMIA, Ministry of Agriculture, Agricultural Research Council, London.

Greb, B. W. 1979. Technology and wheat yields in the Central Great Plains, commercial advances. J. Soil Water Conserv. 34:269-273.

Greb, B. W., D. E. Smika, and J. R Welsh. 1979. Technology and wheat yields in the Central Great Plains, experiment station advances. J. Soil Water Conserv. 34:264-268.

Gupta, S. C., C. A. Onstad, and W. E. Larson. 1979. Predicting the effects of tillage and crop residue management on soil erosion. J. Soil Water Conserv. 34:72-79.

Haas, H. J., C. E. Evans, and E. F. Miles. 1957. Nitrogen and carbon changes on Great Plains soils as influenced by cropping and soil treatments. USDA Tech. Bull. 1164. U.S. Government Printing Office, Washington, D. C.

Jenkinson, D. S. and J. H. Rayner. 1977. The turnover of soil organic matter in some of the Rothamsted classical experiments. Soil Sci. 123:298-305.

Jenny, H. 1930. A study on the influence of climate on the nitrogen and organic matter content of soil. Missouri Agric. Exp. Stn. Res. Bull. 152.

Jenny, H. 1941. Factors of soil formation. McGraw-Hill, New York.

Lal, R., D. DeVleeschawer, and R. Malafa Nganje. 1980. Changes in properties of a newly cleared tropical Alfisol as affected by mulching. Soil Sci. Soc. Am. J. 44:827-833.

Larson, W. E., C. E. Clapp, W. H. Pierre, and Y. B. Morachan. 1972. Effects of increasing amounts of organic residues on continuous corn: II. Organic carbon, nitrogen, phosphorus, and sulfur. Agron. J. 64:204-208.

Larson, W. E., R. F. Holt, and C. W. Carlson. 1978. Residues for soil conservation. pp. 1-15. In W. R. Oschwald (ed.) Crop residue management systems. ASA Spec. Pub. 31. Am. Soc. of Agron., Madison, Wis.

Leeper, R. H., E. C. A. Runge, and W. M. Walker. 1974. Effect of plant-available stored soil moisture on corn yields. I. Constant climatic conditions. Agron. J. 66:723-727.

Lindstrom, M. J., S. C. Gupta, C. A. Onstad, W. E. Larson, and R. F. Holt. 1979. Tillage and crop residue effects on soil erosion in the Corn Belt. J. Soil Water Conserv. 34:80-82.

Massee, T. W. and J. W. Cary. 1978. Potential for reducing evaporation during summer fallow. J. Soil Water Conserv. 33:126-129.

National Fertilizer Institute. 1962. Our land and its care. Washington, D. C.

Onstad, C. A. and M. A. Otterby. 1979. Crop residue effects on runoff. J. Soil Water Conserv. 34:94-96.

Papendick, R. I., J. B. Swan, and W. C. Robocker. 1980. Dryland agriculture in dryland regions. In Proc. Int. Congr. on Dryland Farming, Adelaide, Australia, 25 Aug. - 5 Sept., 1980. (in press).

Parr, J. F. and R. I. Papendick. 1978. Factors affecting the decomposition of crop residues by micro-organisms. p. 101-129. In W.R. Oschwald (ed.), Crop residue management systems. ASA Spec. Pub. 31. Am. Soc. of Agron., Madison, Wis.

Paul, E. A. and J. A. VanVeen. 1978. The use of tracers to determine the dynamic nature of organic matter. Int. Congr. Soil Sci. Trans. 11th (Edmonton, Alberta) III:61-102.

Rasmussen, P. E., R. R. Allmaras, C. R. Rohde, and N. C. Roager, Jr. 1980. Crop residue influences on soil carbon and nitrogen in a wheat - fallow system. Soil Sci. Soc. Am. J. 44:596-600.

Richards, S. J., R. M. Hagen, T. M. McCalla. 1952. Soil temperature and plant growth. pp. 303-480. In B. T. Shaw (ed.) Soil physical conditions and plant growth. Academic Press, New York.

Russell, E. W. 1973. Soil conditions and plant growth. Longman, London.

Skidmore, E. L., M. Kumar, and W. E. Larson. 1979. Crop residue management for wind erosion control in the Great Plains. J. Soil Water Conserv. 34:90-94.

Smika, D. E. and C. J. Whitfield. 1966. Effect of standing stubble on storage of winter precipitation. J. Soil Water Conserv. 21:138-141.

Stevenson, F. J. 1965. Origin and distribution of nitrogen in soil. In W. V. Bartholomew and F. E. Clark (ed.) Soil nitrogen. Agronomy 10:1-42. Am. Soc. of Agron., Madison, Wis.

Triplett, G. B., Jr. and D. M. Van Doren, Jr. 1977. Agriculture without tillage. Sci. Am. 236:28-34.

USDA. 1971. Basic statistics - national inventory of soil and water conservation needs. Stat. Bull. 461. Soil Conservation Service. U. S. Dept. of Agriculture, Washington, D. C.

USDA. 1975. Agricultural Statistics. U. S. Dept. of Agriculture. Washington, D. C.

Van Doren, D. M., Jr. 1965. Influence of plowing, disking, cultivation, previous crop and surface residues on corn yield. Soil Sci. Soc. Am. Proc. 29:595-597.

Van Doren, D.M. Jr. and R.R. Allmaras. 1978. Effect of residue management practices on the soil physical environment, microclimate, and plant growth. pp. 49-83. In W. R. Oschwald (ed.) Crop residue management systems. ASA Spec. Pub. 31. Am. Soc. of Agron., Madison, Wis.

Willis, W. O. and M. Amemiya. 1973. Tillage management practices: Soil temperature effects. pp. 22-42. In Conservation tillage -- The Proc. of a Nat. Conference. Soil Cons. Soc. of Am., Ankeny, Iowa.

Wischmeier, W. H. and D. D. Smith. 1978. Predicting rainfall-erosion losses-a guide to conservation planning. Agric. Handb. No. 537. USDA. U.S. Government Printing Office, Washington, D.C.

John H. Martin, Jr.

7. Biogas Production from Manures: A Realistic Assessment

During the past decade, temporary but serious shortfalls in supplies of natural gas, liquified petroleum gas, gasoline, and distillate fuels have occurred in the United States. In addition, costs for all forms of energy have risen substantially. One consequence of these events has been a growing interest in anaerobic digestion as an energy-producing process using livestock and poultry manures as feedstocks. This paper briefly reviews the history of biogas production from manures, examines the availability of manures as feedstocks, and discusses biogas characteristics and possible uses. The value of digested manures as sources of plant nutrients and feedstuffs is also discussed.

History of Manurial Biogas Production

The concept of using anaerobic digestion to produce energy is not new. As noted by Tietjen (1975), the production of biogas from manures for heating and illumination was suggested by Pasteur in the 1880s. In the 1930s, there were several comprehensive studies of this topic (Buswell and Boruff, 1933; Jacobs, 1934; Buswell and Hatfield, 1936).

During and immediately after World War II, a number of systems for manurial biogas production were built on farms in France and Germany in response to war-related disruptions in fuel supplies (Tietjen, 1975). Use of these systems declined rapidly, however, as supplies of conventional fuels were restored. In 1976, only one of the more than one-thousand digesters built on farms in France and Germany was known to be in operation (Jewell et al., 1976).

In the U.S., a renewal of interest in anaerobic digestion as a process for producing energy from animal manures occurred in the mid-1970s, stimulated by the oil embargo of 1973 and subsequent crude oil price increases. Since 1975, a number of experimental systems have been constructed and studied. Examples include digesters located at Cornell University, Pennsylvania State University, University of Missouri, Washington State Reformatory Honor Farm, and U.S. Meat Animal Research Center. At present, several digesters are in operation on commercial farms.

Availability of Manures as Feedstocks

It has been estimated that the dry matter content of manures excreted by U.S. livestock and poultry exceeds 100 million metric tons per year (Van Dyne and Gilbertson, 1978). Assuming an average energy content of 1.7×10^{10} joules per metric ton of manurial dry matter (Schlesinger et al., 1978), the energy content of U.S. manure production is 1.7×10^{18} joules per year (1.6×10^{15} BTU/yr). This is equal in energy content to about 300 million barrels of crude oil.

However, not all animal manures are collectible and therefore available as feedstocks for biogas production. Van Dyne and Gilbertson estimated that less than half of all the livestock and poultry manures can be economically recovered for use elsewhere, with almost all collectible animal manures produced in confinement facilities. Table 1 compares estimates of the quantities of manure produced annually by U.S. livestock and poultry and of the quantities that are economically recoverable for use elsewhere. Note that dairy and not beef cattle represent the largest single source of collectible animal manure.

Assuming that 50% of the energy content of collectible manures is recoverable as biogas (Huffman, 1978), roughly 4×10^{17} joules of energy could be obtained annually from livestock and poultry manures, or the equivalent of about 60 million barrels of crude oil. This estimate assumes that all collectible animal manures can be used as feedstocks for biogas production and that there are no losses of readily biodegradable compounds during manure collection and handling. However, organic matter losses can be significant when manure accumulations are subject to weathering for extended periods, which for beef feedlots is typically five to six months (Kuo and Jones, 1978). This translates directly into reduced biogas production potential.

Biogas Characteristics

The principal constituents of biogas are methane and carbon dioxide, which together typically comprise about 99% of total gas volume. Hydrogen, hydrogen sulfide, ammonia,

Table 1. Estimates of the quantities of manure produced annually by U.S. livestock and poultry and the quantities of manure that are economically recoverable (Van Dyne and Gilbertson, 1978).

	TOTAL PRODUCTION		ECONOMICALLY RECOVERABLE	
	Dry Weight (M Metric Tons/Yr)[(a)]	% of Total	Dry Weight (M Metric Tons/Yr)[(a)]	% of Total
Beef Cattle				
(Range)	47.3	46.7	1.7	4.1
(Feedlots)	9.5	9.3	9.5	22.6
Dairy Cattle	22.9	22.6	18.5	44.0
Swine	12.1	12.0	5.0	12.0
Sheep	3.4	3.4	1.5	3.7
Laying Hens	3.1	3.0	3.0	7.0
Broilers	1.9	1.9	1.9	4.5
Turkeys	1.1	1.1	0.9	2.1
TOTAL	101.3	100.0	42.0	100.0

[(a)]Million metric tons per year.

nitrogen gas, and water vapor also can be constituents of biogas but are normally present only in trace amounts. Typically, biogas produced from livestock and poultry manures contains 60% methane and 40% carbon dioxide. The methane content can vary, however, with reported values ranging from 50 to 70%. Factors responsible for this variation include substrate composition, method of digester operations, and activity of the methanogenic bacteria.

Biogas containing 60% methane has an energy density of 22.2 MJ/m^3 (600 BTU/Ft^3), which is less than that of other common fuels (Table 2). With removal of carbon dioxide, the energy density of refined biogas (100% methane) is slightly less than that of natural gas due to the absence of heavier hydrocarbons such as ethane and propane, which are commonly present in natural gas.

Although the energy density of biogas can be signifi-

Table 2. Comparison of the energy densities of biogas and methane to natural gas and liquid fuels (Schlesinger et al., 1978).

Fuels	Energy Density
Biogas (60% CH_4), 20°C	22.2 MJ/m^3
Refined Biogas (100% CH_4), 20°C	37.0 MJ/m^3
Natural Gas (79% CH_4, 14% C_2H_6), 20°C	38.3 MJ/m^3
Gasoline	35.6 MJ/L
Fuel Oil, No. 2	40.1 MJ/L
Liquified Petroleum Gas	25.5 MJ/L

Table 3. Effect of compression on the energy density of refined biogas, with comparison to other fuels.

Fuel	Pressure (MPa, Gauge)[a]	Energy Density (MJ/m^3)[b]	Energy Use for Compression (%)[c]
Refined Biogas (100% methane)	0	37	--
	0.7	293	6.5
	1.4	550	7.6
	2.1	818	9.1
	2.8	1,099	10.2
	3.4	1,380	11.2
Liquified Petroleum Gas	2	25,500	--
Fuel Oil, No. 2	--	40,100	--

[a] 1 MPa = 10^6 Pascals = 9.87 atmospheres
[b] 1 MJ/m^3 = $10^6 J/m^3$ = 26.8 BTU/Ft^3
[c] % of energy available before compression (Smith et al., 1977)

cantly increased by carbon dioxide removal and compression (Table 3), it still compares poorly to liquid fuels. In addition, the energy requirements for compression are significant. To achieve an energy density comparable to gasoline or diesel fuel, liquification of refined biogas is necessary. However, unlike propane and butane, which are the principal constituents of liquified petroleum gas, methane is difficult to liquify, having a critical temperature of -82.4°C (Katz et al., 1959). Above this temperature, methane can not be liquified even with extremely high pressures. The critical temperatures for both propane and butane are significantly higher (Table 4). In addition, almost all impurities must be removed before the methane component of biogas can be liquified. Typical requirements are 0.002% carbon dioxide, 0.0001% water, and 2.3 x 10^{-5} mg of hydrogen sulfide per m^3 of gas (Kohl and Risenfeld, 1974). To achieve this level of purification, molecular sieves would probably be necessary.

Biogas Utilization

Although the technical feasibility of producing biogas from animal manures has been firmly established, the economic feasibility remains unclear. The principal source of this uncertainty is the question of effective biogas utilization. Assuming conventional fuels are available, biogas must provide a reasonable rate of return to invested capital, management, and labor. Otherwise, other investment opportunities become more attractive.

The value of biogas depends on the quantity utilized and the conventional energy source that is replaced. As shown in Table 5, prices for conventional sources of energy, and therefore the value of biogas, vary substantially, ranging from \$2.77/GJ for coal to \$12.50/GJ for electricity (1 GJ = 10^9J = 0.95 x 10^6 BTU).

Because of the low energy density of biogas, as noted earlier (Tables 2 and 3), utilization options are limited. Use of biogas as a fuel for trucks or tractors is possible but is generally viewed as not practical because of the frequent need for refueling. Even with carbon dioxide removal and compression, large fuel tanks would permit only one to two hours of small truck or tractor operation. Low energy density also makes storage of biogas for extended periods impractical. Thus, biogas utilization alternatives are realistically limited to use as a boiler fuel or for generating electricity. Sale of purified biogas as substitute natural gas (SNG) also is a possible option.

Table 4. Critical properties of methane, propane, butane, and carbon dioxide (Katz et al., 1959).

Compound	Critical Temperature (°C)[a]	Critical Pressure (MPa Gauge)[b]
Methane (CH_4)	-82.4	4.6
Propane (C_3H_8)	93.5	4.3
Butane (C_4H_{10})	152.0	3.8
Carbon dioxide (CO_2)	31.0	7.4

[a] The temperature above which a gas cannot be liquified at any pressure.
[b] The pressure necessary to liquify a gas at its critical temperature.

Table 5. Energy densities and prices of conventional energy sources.

Energy Source	Energy Density[a]	Price[b]	Price ($/GJ)
Gasoline	35.6 MJ/L	$0.336/L	9.44
Diesel Fuel	40.1 MJ/L	0.340/L	8.48
Fuel Oil, No. 2	40.1 MJ/L	0.325/L	8.10
Liquified Petroleum Gas	25.5 MJ/L	0.198/L	7.76
Natural Gas	36.3 MJ/m^3	0.155/m^3	4.26
Electricity	3.6 MJ/KWH	0.045/KWH	12.50
Coal, Anthracite	30.3 MJ/kg	0.109/kg	3.60

[a] Schlesinger et al., 1978.
[b] Consumer prices excluding taxes, Ithaca, New York, January 1981.

Of these alternatives, use of biogas as a boiler fuel is the most attractive, particularly if it replaces fuel oil or liquified petroleum gas, due to the high cost of these fuels (Table 5). However, the need for boiler fuels on farms and feedlots is limited. One analysis of energy use on a 100 cow dairy farm indicated that only 21% of annual biogas production could be used as a boiler fuel, primarily for residential space heating and heating of water (Jewell et al., 1976). Thus, there would be a substantial seasonal variation in demand. For a 30,000 head beef feedlot, Sweeten (1979) estimated that only 30% of the biogas produced could be used as a boiler fuel, principally for steam flaking of corn or sorghum.

Due to the limited demand for boiler fuels on farms and feedlots, on-site generation of electricity is frequently suggested for biogas utilization, an option which is attractive for several reasons. One is the ability to use engine-generator set waste heat for digester heating. This increases the net output of biogas and consequently reduces unit production costs. When biogas is used as a boiler fuel for digester heating, as much as 30% of total biogas production may be required. (Actual requirements depend on digester operating temperature and climate.) In addition, on-farm demand for electricity is relatively constant on a seasonal basis, in contrast to space heating requirements.

The thermal efficiency of converting biogas to electricity, however, is poor. Assuming a thermal efficiency of 16% for a biogas-fueled internal combustion engine (Neyeloff and Gunkel, 1975) and an efficiency of 90.9% for a synchronous generator (LaPierre and Jones, 1978), the overall efficiency for converting biogas to electricity is only 14.5%. Although waste heat utilization in addition to digester heating would improve overall thermal efficiency, analyses by Jewell et al. (1976) and Smith et al. (1977) have shown that waste heat availability and demand are typically six months out of phase.

The economic attractiveness of using biogas to generate electricity depends on local rates. If electricity can be purchased at $12.50/GJ ($0.045/KWH), as in Table 5, the combined cost of biogas production and electricity generation cannot exceed $1.81/GJ of biogas. Lizdas et al. (1980) have estimated a biogas production cost of $5.97/GJ for an 80,000 head beef feedlot. For a 100 cow dairy farm with a simple plug flow digester, biogas production costs have been estimated to be only $3.07/GJ (Jewell et al., 1980). However, neither of these estimates included costs of generating electricity. Thus, the cost of electricity purchased from

Table 6. Comparison of plant nutrient concentrations in livestock and poultry manures before and after anaerobic digestion (g/L).

		Nitrogen	Phosphorus	Potassium
Dairy[a]	Influent	2.6	0.6	1.9
	Effluent	2.9	0.5	2.0
Beef[b]	Influent	4.3	--	--
	Effluent	3.9	--	--
Swine[c]	Influent	3.9	1.3	1.2
	Effluent	4.1	1.2	1.2
Laying Hen[d]	Influent	9.3	3.5	3.6
	Effluent	9.6	2.5	3.9

[a]Converse, Graves, and Evans (1977).
[b]Hashimoto, Chen, and Prior (1978a).
[c]Fischer et al. (1978).
[d]Converse et al. (1981).

Table 7. Comparison of the nutrient composition of beef cattle manure before and after anaerobic digestion (g/L). (Prior and Hashimoto, 1981).

Component	Influent	Effluent	% Change
Dry Matter	72.2	39.8	-44.9
Crude Protein	25.1	24.5	- 2.4
Amino Acids	10.3	9.4	- 8.7
Cellulose	7.6	4.2	-44.7
Hemicellulose	18.9	8.0	-57.7
Lignin	2.2	2.6	+18.2
Ash	7.1	6.8	- 4.2
Gross Energy (Mcal/L)	336.5	185.3	-44.9

a public utility must be substantially higher than \$12.50/GJ or biogas production costs must be substantially lower than those noted above for this utilization option to be economically attractive.

Byproduct Recovery

As shown by Van Dyne and Gilbertson (1978), collectible livestock and poultry manures contain substantial quantities of nitrogen, phosphorus, and potassium that can be used for crop production, thereby replacing manufactured fertilizer materials. One particularly attractive aspect of anaerobic digestion is that it conserves these plant nutrients (Table 6). In contrast, other processes to obtain energy from manures, such as direct combustion, pyrolysis, and partial oxidation, result in an almost complete loss of nitrogen. It should be noted, however, that anaerobic digestion does not enhance the value of manures as sources of plant nutrients. Thus, the monetary value of anaerobically digested manures as sources of plant nutrients is no greater than that of undigested manures.

Several investigators, including Burford and Varani (1978), Hashimoto, Chen, and Prior (1978b), and Lizdas et al. (1980), have suggested that anaerobic digestion enhances the nutritive value of manures, particularly beef cattle manures, as feedstuffs. This provides the source of a substantial by-product credit to the biogas production process. Assuming that undigested beef cattle manure has no nutritive or monetary value and that digested manure is comparable to cottonseed meal, Lizdas et al. (1980) have suggested that biogas can be produced at a profit of \$19.91/GJ based on the difference between the estimated feedstuff value of the digested manure and production costs. The estimated cost of biogas production without use of digested manure as a feedstuff was \$5.97/GJ.

However, these estimates should be looked at very critically. Comparison of the nutrient composition of beef cattle manure before and after anaerobic digestion (Table 7) indicates that anaerobic digestion does not enhance, but rather decreases, the value of manure as a feedstuff. The substantial decreases in digestive fiber, cellulose and hemicellulose, and gross energy are particularly noteworthy. Measurements of digestibility and results of feeding trials have shown that anaerobically digested beef cattle manure has little value as a feedstuff for ruminants and can decrease animal performance when used as a component of finishing rations for beef cattle (Martin, Pilbeam, and Loehr, 1980).

Summary

Although the use of anaerobic digestion to produce energy from animal manures is both attractive in principal and technically feasible, biogas utilization options are limited. Biogas can not realistically be used as a fuel for trucks or tractors. Thus, available options are limited to use as a boiler fuel or for generating electricity. Demand for boiler fuels on farms and feedlots is limited, however, and the cost of producing electricity using biogas is, at best, marginally competitive compared to an average cost for electricity of \$12.50/GJ (\$0.045/KWH). Because of these constraints, there should be a careful analysis of opportunities for energy conservation, options for biogas utilization, and of the costs of the conventional energy sources that could be replaced before biogas production is undertaken.

References

Burford, J. and F. T. Varani. 1978. The Lamar bioconversion plant design. In J. M. Sweeten (ed.), Methane Production from Livestock Manure, pp. 77-101. Texas Agricultural Extension Service, Texas A&M University, College Station.

Buswell, A. M. and C. S. Boruff. 1933. Mechanical equipment for continuous fermentation of fibrous materials. Industrial and Engineering Chemistry 25:147.

Buswell, A. M. and W. D. Hatfield. 1936. Anaerobic Fermentations. State Water Survey Bulletin 32. Urbana, IL.

Converse, J. C., R. E. Graves, and G. W. Evans. 1977. Anaerobic digestion of dairy manure under mesophilic and thermophilic temperatures. Transactions of the American Society of Agricultural Engineers 20(2):336-340.

Converse, J. C., G. W. Evans, K. L. Robinson, W. Gibbons, and M. Gibbons. 1981. Methane production from a large-size on-farm digester for poultry manure. In Livestock Waste: A Renewable Resource, pp. 122-125. American Society of Agricultural Engineers, St. Joseph, MI.

Fischer, J. R., E. L. Iannotti, D. M. Sievers, C. D. Fulhage, and N. F. Meador. 1978. Methane production systems for swine manure. In J. M. Sweeten (ed.), Methane Production from Livestock Manure, pp. 45-76. Texas Agricultural Extension Service, Texas A & M University, College Station.

Hashimoto, A. G., Y. R. Chen, and R. L. Prior. 1978a. Methane and biomass production systems for beef cattle manure. In J. M. Sweeten (ed.), Methane Production from Livestock Manure, pp. 20-44. Texas Agricultural Extension Service, Texas A & M University, College Station.

Hashimoto, A. G., Y. R. Chen, and R. L. Prior. 1978b. Thermophilic anaerobic fermentation of beef cattle residue. In D. Klass and W. Waterman (eds.), Energy from Biomass and Waste, pp. 379-402. Institute of Gas Technology, Chicago, Illinois.

Huffman, W. J. 1978. Alternative manure recycling systems for energy recovery. In J. M. Sweeten (ed.), Methane Production from Livestock Manure, pp. 3-19. Texas Agricultural Extension Service, Texas A & M University, College Station, Texas.

Jacobs, P. B. 1934. Heat and light on the farm--the utilization of farm wastes. Architectural Record 7:358.

Jewell, W. J., H. R. Davis, W. W. Gunkel, D. J. Lathwell, J. H. Martin, Jr., T. R. McCarthy, G. R. Morris, D. R. Price, and D. W. Williams. 1976. Bioconversion of Agricultural Wastes for Pollution Control and Energy Conservation. Report No. TID-27164. Energy Research and Development Administration, Washington, D.C.

Jewell, W. J., S. Dell'Orto, K. J. Fonfoni, T. D. Hayes, A. P. Leuschner, and D. F. Sherman. 1980. Anaerobic Fermentation of Agricultural Residue: Potential for Improvement and Implementation, Vol. II. Report No. DOE/ET/20051-T1. U.S. Dept. of Energy, Washington, D.C.

Katz, P. L., D. Cornell, R. Kobayashi, F. H. Poettmann, J. A. Vary, J. R. Elenbaas, and C. F. Weinaug. 1959. Handbook of Natural Gas Engineering. McGraw-Hill Book Company, New York.

Kohl, A. and F. Risenfeld. 1974. Gas Purification, 2nd edition. Gulf Publishing Company, Houston, Texas.

Kuo, M. C. T. and J. L. Jones. 1978. Environmental and energy output analysis for the bioconversion of agricultural residues to methane. In D. Klass and W. Waterman (eds.), Energy from Biomass and Waste, pp. 143-175. Institute of Gas Technology, Chicago, IL.

LaPierre, W. A. and B. M. Jones. 1978. Electrical and electronics engineering. In Marks' Standard Handbook for Mechanical Engineers, 8th edition, pp. 15.1-15.97. McGraw-Hill Book Company, New York.

Lizdas, D. J., W. B. Coe, M. Turk, and F. S. Fowler. 1980. Methane Generation from Cattle Residue at a Dirt Feedlot. Report No. DOE/ET/20039-20. U.S. Dept. of Energy, Washington, D.C.

Martin, J. H., Jr., T. E. Pilbeam, and R. C. Loehr. 1980. Nutritive value of anaerobically digested beef cattle manure as a feedstuff. Paper No. 80-4546. American Society of Agricultural Engineers, St. Joseph, MI.

Neyeloff, S. and W. W. Gunkel. 1975. Methane-carbon dioxide mixtures in an internal combustion engine. In W. J. Jewell (ed.), Energy, Agriculture, and Waste Management, pp. 397-408. Ann Arbor Science, Ann Arbor, MI.

Prior, R. L. and A. G. Hashimoto. 1981. Potential for fermented cattle residue as a feed ingredient for livestock. In D. L. Wise (ed.), Fuel Gas Production from Biomass. CRC Press, Boca Raton, FL.

Schlesinger, M. D., J. F. Farnsworth, M. H. Mawhinney, C. R. Velzy, C. O. Velzy, and W. E. Lewis. 1978. Fuels and furnaces. In Marks' Standard Handbook for Mechanical Engineers, 8th edition, pp. 7.1-7.69. McGraw-Hill Book Company, New York.

Smith, R. J., R. L. Fehr, J. A. Miranowski, and E. R. Pidgeon. 1977. The role of an anaerobic digester on a typical central Iowa farm. In R. C. Loehr (ed.), Food, Fertilizer, and Agricultural Residues, pp. 341-371. Ann Arbor Science, Ann Arbor, MI.

Sweeten, J. M. 1980. Energy recovery from feedlot manure: an assessment of alternate technologies. Texas Agricultural Extension Service, Texas A & M University, College Station.

Tietjen, C. 1975. From biodung to biogas--historical review of European experience. In W. J. Jewell (ed.), Energy, Agriculture, and Waste Management, pp. 247-259. Ann Arbor Science, Ann Arbor, MI.

Van Dyne, D. L. and C. B. Gilbertson. 1978. Estimating U.S. livestock and poultry manure and nutrient production. Report ESCS-12, Economics, Statistics, and Cooperatives Service, U.S. Department of Agriculture, Washington, D.C.

William Lockeretz

8. Concluding Comments: New Problems Demand Old Solutions

In these concluding pages, claiming the editor's prerogative of having the last word, I offer some views on the right and the wrong ways of solving our agricultural energy problems. Unlike the objective scientific analyses comprising the rest of this book, this chapter consists of purely personal opinions. They are, however, based on several years of very close, first-hand observation. Readers who have been involved in this field in other ways may find that the characterizations don't tally at all with their own experiences. But I have tried to describe what I believe to be a large and increasingly important category of agricultural/energy activities, whether on the part of government, industry, consulting companies, national laboratories, or universities.

A distressing feature of a lot of what is being written in these circles about energy and agriculture is the rather free use of some fairly strong words, like "crisis" and "post-petroleum era." Such words certainly help create a sense of urgency, at least the first dozen or so times you hear them. But they also may reflect a serious shortcoming in the current approaches to the agricultural energy situation, a shortcoming beyond their merely being a contribution to the general devaluation of words. The deeper shortcoming is a lack of a sense of history or of the continuity of intellectual and technical evolution.

It is a commonplace about today's times that there never has been such rapid change, not only in technology, but also in political and social institutions, and even in ways of thinking. ("We have to teach children the new ways of thinking they will need to survive in the 21st century.") Like all commonplaces, it has very limited value. In fact, it could even be wrong--I offer 1775-1800 and 1914-1939 as periods of as much change as the past quarter century, or the 1940s as a decade of as much change as the 1970s. But even if today's conditions are fundamentally different from anything that came before, that doesn't automatically mean--as some people working in agriculture and energy apparently believe--that we need fundamentally different ways of analyzing our problems and doing something about them. Some new technology may be needed, for sure. Some new institutional arrangements may be needed to permit these technologies to flourish. But an obsession with newness and the future should not overshadow the enormous resource represented by the decades of continuous progress in analyzing earlier social and technical problems. Appreciating this resource can provide us with most of what we need to deal with our current problems, even if some of the details must be worked out afresh.

Much of the research, development, and demonstration work in agriculture and energy has a transient and ephemeral quality about it. In part this is a political phenomenon. Because the money most often comes from the government, and because energy is a more political issue than many other technical and scientific topics, work in this field is very vulnerable to short-term changes in the political climate (lasting around four years, say).

But part of the problem is found in the intellectual outlook of the individuals doing the work--it can't all be blamed on politicians and bureaucrats. I have heard impatient researchers say "It's taking so long to get my report out that by the time it's published it'll be out of date." Undoubtedly such a person takes great pride in being very current and up-to-the-minute. But to me it's a very serious criticism of one's own work--can you really be proud of something that will have no value a few months from now? The corollary is to believe that nothing written more than a short while ago can have much value for you. I was struck by a report on a research project on methane production from organic wastes that began by reviewing the rather extensive and successful European experience in building and operating full-size commercial plants in the 1930s. Apparently the authors realized that the reader might wonder why the current project was necessary. The explanation: the European experience was from the 1930s.

The intellectual conceit that lets someone ignore the past generally also leads to ignoring even one's contemporaries. I have sat through so many presentations on agricultural energy research and demonstration projects--generally at contractor review meetings intended to increase coordination among related efforts--in which each project is described as though it were the only one of its kind anywhere, ever. There is no sense of the field being a cumulative one. There is no sense of an obligation to explain what gap, if any, in the world's knowledge the project was intended to fill. Rather, building the particular piece of apparatus is an end in itself, an activity that needs no further justification as far as its value to anyone else working in the same area. The consequence is that the measure of the quality of such an effort is the technical sophistication of the set-up, not the usefulness of the information acquired. And no matter that someone else may already have worked out a simpler, cheaper, or more reliable way of getting the same job done.

The papers in this book have, I hope, avoided these shortcomings. Even when the authors are presenting their own new research, as well as when reviewing previous work, the analyses are intended to fit into an ongoing body of activity. Where appropriate, the authors then go on to suggest what we need to do in their specific areas. I would like to conclude with some more general ideas about what we need.

Actually, they are ideas about what we don't need.

- We don't need new, interdisciplinary, systems analysis methodologies for analyzing our agriculture and energy problems;
- We don't need new channels of communication to get results into the hands of users more quickly;
- We don't need greater flexibility in how our agricultural and energy agencies are organized so that they can adjust to constantly changing demands and problems;
- We don't need to develop more sophisticated planning techniques to design more effective programs and strategies for new energy technologies;
- We don't need to get the answers quickly.

Good research on agriculture, on energy, indeed on most applied scientific and technical problems, has always taken a "systems analysis" approach, if that just means looking at how the components of some system fit together, rather than merely examining them in isolation. No matter that the terms are relatively recent--the idea isn't. But the terms have been elevated to a status they don't deserve (and a length they don't need--but why say "method" when "methodology" will do?) Terms like "systems analysis" have become hardly more than labels, bits of jargon to make clear to those who might not have realized it that the speaker is "state of the art." By having been separated from content to an extreme degree, they have been rendered devoid of intellectual content. "Development of a conceptual systems methodology" to solve some problem is fine, but solving it is even better. Being trained in a "systems approach" does not exempt a person from needing a thorough familiarity with the particular system under consideration. (I remember reading one report where I got the distinct impression that the author did not realize that the corn that we grow 70 million acres of is not the kind you eat with butter and salt.) This is not to excuse the extreme overspecialization that has characterized much former research and training, especially in agriculture. But the best researchers--those who have surmounted the limitations of overspecialization--have no reason to fear that their old-fashioned training and experience will prove inadequate for the unprecedented complexity of our current agricultural and energy situation.

Of course anyone who has developed a new "methodology" would naturally be eager to see it in print. It is frustrating to have to wait around, perhaps even at the risk that some even more unprecedented situation requires the development of an even newer methodology. Thus the (unprecedented) volume of self-published, privately distributed reports. But there is still a lot to be said for the old-fashioned refereed journal, and even for books. Whatever its problems, the referee system offers at least some minimal quality control, and the economics of commercial book and journal publication keeps the volume of material down to something almost manageable. When these constraints are removed, we know for sure only that the total volume of printed material increases. It is not clear that the total value of it increases proportionally, or even at all. Also, under the traditional system, whatever manages to get published is more accessible and stays around longer. If you have access to a good research library, or even to a library that gets inter-library loans, you can get an article that was published five, twenty, or fifty years ago. It doesn't depend on whether the self-publishers happened to have printed enough to meet the demand, or whether they even are still in existence by the time you are interested in the article. (See later discussion of bureaucratic reorganization.) Of course, journal publication is slower. That is probably its single greatest advantage. If something is not worthwhile by the time it gets into print, maybe it wasn't worth much the day it was written. (We are talking about science, after all, not stock market quotations or race results.) The intolerably tedious process of revising a report in accordance with someone else's suggestions gives the material a needed opportunity to ferment, and the knowledge that what you write is supposed to sound reasonable as much as a few months later should exert a beneficial restraint on how eager you are to subject your latest work to immediate public scrutiny by committing it to print. Indeed, it might even raise a question as to whether it should be committed to print at all.

Like self-publication, periodic reorganization of government agencies is nothing new, of course. But we are living in an era--I am tempted to claim--when agency reorganizations are occurring at a rate unprecedented in the world's history. (After all, the Department of Energy has been reorganized more times in my lifetime than in the previous thousand years.) I don't think I am alone in the experience of having a proposal solicited by one bureaucratic entity, then evaluated by its successor, and finally having its performance monitored by a third version. Such an arrangement is not very conducive to efficient performance on the part of a research contractor, no matter how glowingly the agency describes the benefits that it expects the new organization to have on its own performance. But a change is not necessarily a difference, and a difference is not necessarily an improvement, certainly not when it's done mainly to give an appearance of doing something. Invariably, the same people end up filling the boxes on the newly drawn organization chart, since they are the only ones with the needed experience. (Of course, the fact that a reorganization was considered necessary suggests that maybe some other kind of experience is what was really needed, but it takes an outsider to point this out.) Working with an agency whose own time horizon is so short helps shorten the time horizon of the researcher's own thinking, and helps create the feeling that one's work only has to be valuable for the next few months.

It also causes trouble with program planning. There is nothing wrong with planning, in principle. But you have a real problem when the scope of the plan exceeds the lifetime of the agency that is supposed to follow the plan. Consider, for example, the situation of a well-intentioned staff member of the Federal Energy Administration who might have tried to formulate a five-year plan back in 1976. He probably never would have guessed that before the period was even half over, not only would the agency have been replaced, but also that the replacement would have been replaced. Likewise, in 1980, the replacement's replacement, the Department of Energy, directed certain institutions that it supported to formulate program plans for 1983-87. But as this is being written (mid-1981), it is not at all clear that those institutions will be allowed to live to see 1982. And of course, while these directions were being worked out, which happened to be around the time of the 1980 presidential campaign, the Department of Energy was itself the subject of certain less-than-five-year plans on the part of one of the candidates, who announced his plans to abolish it. (That candidate now has the Department of Energy as part of his administration.)

Finally, do we really need quick answers? This discussion suggests that the perceived need for quick answers, and researchers' willingness to try to provide them, comes about not from the changing nature of our agricultural and energy problems, but rather from the changing structure of the business of dealing with such problems. Research does not necessarily proceed at a rate determined by internal logic, that is, by the current state of development of a particular field and the rate at which advances are thought possible. Instead, the rate is determined largely by the economic, political, and bureaucratic needs of the people who support the work and the people who carry it out. The traditional structure of publicly supported research, which is based largely on people with a long-range involvement in a subject and who have a major role in deciding what areas would be worked on and over what time-scale the work would be done, apparently does not meet these needs. Thus if some agency wants to be able to go before a Congressional committee and show it a program that particularly benefits an important member's home district, it needs a research community that can deliver the goods at the right time. The agency doesn't support the research--the research supports the agency. This means that research must be available on a fee-for-service basis, and as a result we have the knowledge-for-sale industry that treats research as just another article of commerce. The new kind of research operation that has developed to meet this need naturally has to have people who can work on whatever new business opportunity comes along, not just those that they might have chosen to work on in line with their interests and qualifications. "Systems analysis" supplies the theoretical basis for such an enterprise. If systems analysis transcends the details of the particular system being analyzed, then you need only make sure that you hire people who understand "systems" and who can "see the big picture," and they can work on anything. And since systems analysis teaches them to know an irrelevant empirical or descriptive detail when they see it, they can do the work quickly. So quickly that they can be safely involved in some other work by the time anyone else has had a chance to discover that maybe the systems approach wasn't quite enough to handle the problem.

Occasionally, some problem may really need a basically new approach. If this approach turns out to be successful, it might possibly be a good idea to let other people know about it directly, without first letting anyone else have a chance to evaluate it. Eventually, an agency might do well to reorganize itself to reflect the new insights generated by such an approach, and to develop new program plans to put them into action. And it might even be important to be able to do all this as quickly as possible.

But let's not overdo it. Agriculture's problems didn't begin yesterday, as any farmer can tell you, and they'll be around tomorrow. Our energy problems didn't begin yesterday either. In fact, they didn't even begin on the day of the oil embargo. If a "crisis" is really a "crisis," and if we really are entering the "post-petroleum era," then the right way to think about energy shouldn't change much if the price of gasoline happens to go up or down a few cents a gallon one day, or if the makeup of Congress changes in one election. To think over a slightly longer time horizon wouldn't hurt. Of course, believing that the world has never changed so rapidly, so that its problems have never been so challenging and complex, may be flattering and self-aggrandizing for those who are supposed to be doing something about them. But implicity, this belief deprecates those who were already working on these issues before whatever happens to be the current research fashion caught on. And it deprives us of the great benefit of being able to build on what has already been done. The pursuit of the latest version of what is chic or trendy is harmless enough for things like clothes and talk shows. But if we really believe that the agriculture-energy issue is as urgent and critical as it invariably is made out to be when public research money is being sought, then in return the public is entitled to something a little more durable.

Index